AF766137

THE FRONT LINE

IMAGES OF NEW ZEALANDERS
IN THE SECOND WORLD WAR

THE FRONT LINE

GLYN HARPER
with Susan Lemish

MASSEY UNIVERSITY PRESS

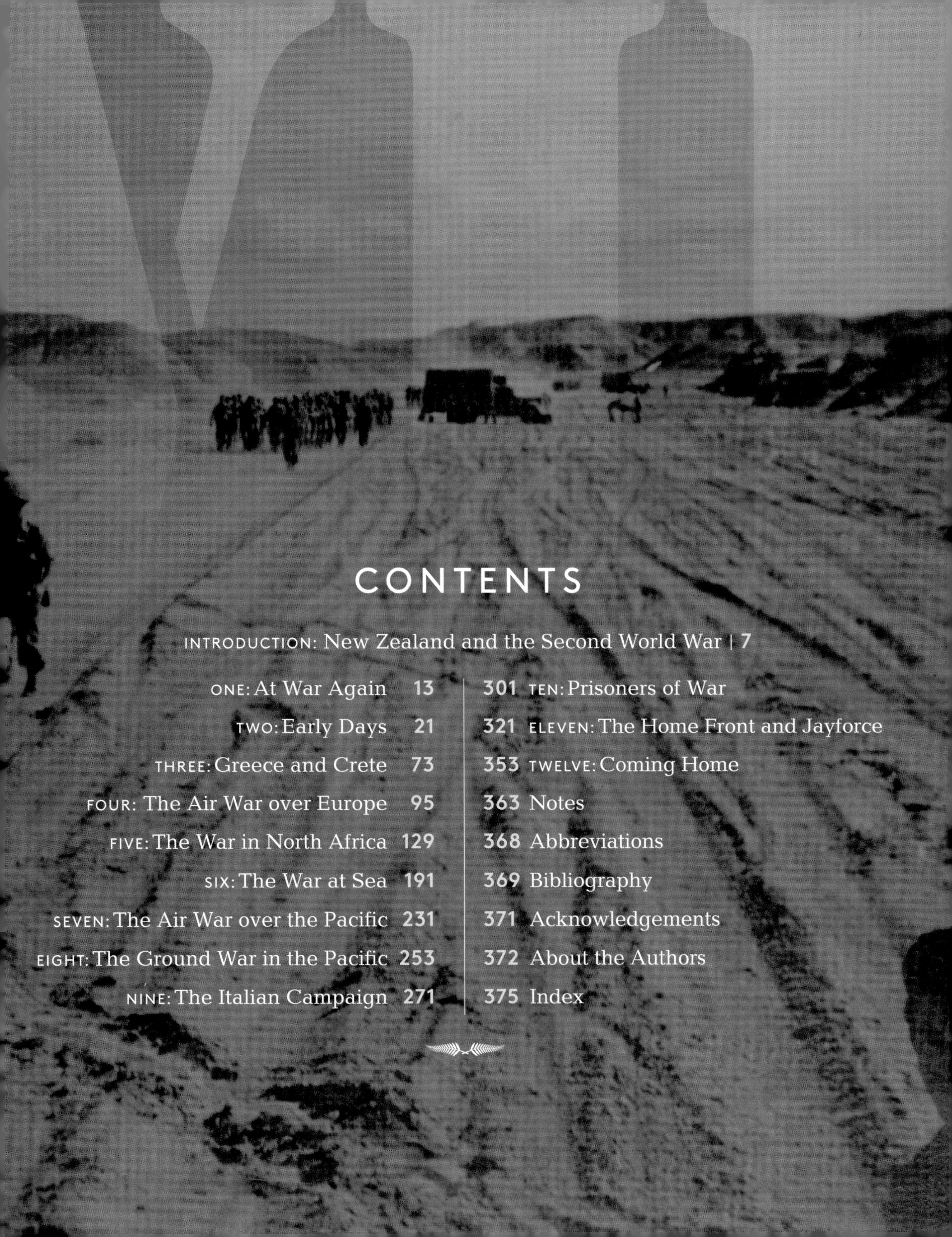

CONTENTS

ABOVE | A medic offers water while checking on a wounded soldier in North Africa.

OZ TURLEY, 6TH FIELD REGIMENT, FROM TONY GOODWIN

New Zealand and the Second World War

The war of 1939–45 remains the bloodiest conflict in human history. Even 80 years later, historians cannot be sure of the extent of loss of life,[1] though it is usually estimated as somewhere between 50 and 60 million people. In his impressive and appropriately titled history *All Hell Let Loose* Max Hastings suggests that 'at least' 60 million were killed, which was an average of 27,000 people for each day of the war between September 1939 and August 1945.[2] One thing is certain, though. This was truly a global struggle: hundreds of millions of the world's inhabitants had their ordinary, peaceful lives shattered and were plunged into the ordeals and terrors of the most comprehensive war in history. New Zealand, despite its distance from the centre of events, was caught up in the whirlwind from the beginning. As James Belich has acknowledged, 'New Zealand's role in World War Two was less lethal than in World War One, but just as traumatic.'[3]

In 1939 New Zealand had a population of just over 1,600,000 and the country was still feeling the effects of the great economic depression that had haunted much of the preceding decade. When war broke out there were still 9000 people unemployed, with a further 22,000 on subsidised government work schemes.[4] Despite these problems, New Zealand was an early entrant into the Second World War and, of the democratic nations that participated, fought for the longest period, along with Australia and Britain. The conflict is now generally accepted to have lasted 2179 days, from the German invasion of Poland on 1 September 1939 to the surrender of Japan on 15 August 1945. New Zealand was at war for all but three of those days.[5]

Despite its small population, New Zealand's war effort was massive. More than 65 per cent of all men of military age were mobilised, and some 205,000 men and women, almost one in eight New Zealanders, served in the armed forces.[6] In addition, though, more than 250,000 New Zealanders served in the Home Guard or various other reserve and emergency forces. As well as the 10,000 New Zealand women who joined the armed forces, a further 75,000 served with the Women's War Service Auxiliary and 2700 with the Women's Land Service. This level of

PAGE 2 | Members of the Women's Royal New Zealand Naval Service (Wrens) parade down an Auckland street in February 1945.

NATIONAL MUSEUM OF THE ROYAL NEW ZEALAND NAVY, APD0108

PAGES 4–5 | New Zealand troops on a route march near Maadi Camp, Egypt.

NATIONAL ARMY MUSEUM, 1990-1010

human mobilisation was matched by the effect on the New Zealand economy: war-related expenditure consumed more than half of the national income between 1942 and 1944.[7]

Of those New Zealanders who joined the armed services during the war, most served in frontline combat units. The Second New Zealand Expeditionary Force (2 NZEF) was concentrated around a 10-battalion infantry division. This, later named 2 New Zealand Division, experienced much hard fighting and was part of military disasters in Greece, Crete, North Africa and Italy. A two-brigade infantry division, the 3rd New Zealand Division, took part in several amphibious landings in the Pacific in 1943 before manpower shortages forced its dissolution in late 1944. In the air, New Zealand fully supported the Empire Air Training Scheme (EATS), which trained aircrew in Canada for the Royal Air Force (RAF). As a result, New Zealanders flew in every theatre of war. At sea, much of New Zealand's contribution was made by providing personnel for the Royal Navy, which had a worldwide reach.

Wherever New Zealand soldiers fought in the war, nurses of the New Zealand Army Nursing Service (NZANS) were there to care for them. By May 1940, more than 1200 nurses had volunteered for overseas service. Between 1939 and 1945, 602 members of the NZANS worked abroad; many New Zealand nurses served with other medical organisations.[8]

The cost of such a heavy commitment was bound to be high. Of those who served, some 40,000 were killed, wounded or taken prisoner.[9] The 11,671 members of the New Zealand armed forces and Merchant Navy who died during the war is a proportionately higher loss (per capita) than that suffered by Britain, and twice that of Australia — and neither of those two countries had an easy war.[10]

As THE LATE Richard Holmes observed, the Second World War was 'a photographer's war. Although there is abundant film . . . somehow it is the photograph that freezes the moment for posterity'.[11] Conflicts had been recorded on camera since the Mexican–American War of 1846 and reached a state of maturity during the First World War, but photography came of age during the Second World War. Some images are justly famous, capturing a moment in time for posterity: St Paul's Cathedral standing amid the smoke of the London Blitz, British Prime Minister Winston Churchill with a Tommy gun giving his V for victory sign, Australian infantry attacking through the smoke and dust of El Alamein, laden infantrymen wading ashore on the beaches of Normandy, the raising of the US flag at Iwo Jima. Unfortunately, as we now know, some of these — the Iwo Jima photograph, the Australians at El Alamein, and Douglas MacArthur wading ashore through the Philippines surf — were staged shots or re-enactments of earlier events. The fact that they were made at all, however, points to the power of the photograph in conveying an experience that we want to believe is true. That said, this book has tried to avoid using images that are not obviously candid and which are immediate.

ABOVE | In May 1940, the Second Echelon departed from New Zealand. This troopship, the *Andes*, is departing from the South Island. Unlike the First Echelon, which sailed to Egypt to be formed into the 4th Infantry Brigade and Divisional Headquarters, the Second Echelon would disembark in the United Kingdom, which was then facing the threat of invasion. While in the United Kingdom it would be formed into the 5th Infantry Brigade.

NATIONAL MUSEUM OF THE ROYAL NEW ZEALAND NAVY, OC 628

Most of the photographs used are from the magnificent collections of New Zealand's three service museums: the National Army Museum in Waiouru; the National Museum of the Royal New Zealand Navy at Torpedo Bay, Auckland; and the Air Force Museum of New Zealand at Wigram in Christchurch. A February 2019 appeal to the public to share their Second World War photographs produced an outstanding response; some of those images are included. The majority of the photographs in this book are published here for the first time.

In compiling the book, consideration has been given to include every battle and theatre in which New Zealanders fought. Also represented is the home front, which was a vital part of New Zealand's contribution to the Allied war effort. Live action shots are the gold standard of war photography but, as with the First World War, they remain incredibly rare — and those that do exist are frequently published. Such images have been used wherever possible and only when their veracity had been firmly established.

There were some surprises. One revelation was just how many German photographs appeared in soldiers' albums. These must have been developed from cameras 'ratted' from German prisoners of war or taken from the dead. Many German soldiers carried and used cameras, which were of the highest quality. Why New Zealand soldiers took the effort to develop the film and put the images in their albums is a puzzle, and a topic worthy of further investigation. Another discovery was the number of photographs taken by New Zealand prisoners of war. This was surprising as cameras were generally banned in prison camps.

All the captions have been written using the original annotations for guidance. In far too many cases, no information about an image was available and guesswork would have been required. Where there was no information about a photograph, it was not used.

Rather than organising the images chronologically, which could have been a little plodding, they have been grouped around campaigns and themes. It is hoped that this will sustain interest, avoid artificial breaks when campaigns crossed yearly boundaries, and help readers to home in on the sections that are of most interest to them and their families. There is a brief outline of each campaign or theme, to place the photographs in the context of both the war's progress and New Zealand's contribution.

In early September 1939 a future New Zealand major-general named Howard Kippenberger listened to British Prime Minister Neville Chamberlain announce, on the radio, that a state of war now existed between Britain and Germany. He turned to his young son and wife, whose face had gone white, and said, 'That's the end of the old pleasant days.'[12] While raising and training his battalion in Tai Tapu, outside Christchurch, Kippenberger, along with his infantry soldiers, heard 'Now Is the Hour' for the first time at a campfire concert. This haunting song of

farewell, also known as 'Po Atarau', was often played at the wharf as troopships left New Zealand. Kippenberger and his men felt 'for a moment the cold hand of fate and the shadow of the long years ahead'.[13]

Like any war, this one was strewn with moral complexities. The strategic bombing of Germany and the use of atomic weapons against Japan continue to cause concern and are hotly debated topics. Many people, however, still regard the 1939–45 conflict as a just war. In the words of Richard Holmes, 'Yet ultimately this was a war in which good was pitted against evil; and if the world which emerged from it brought tensions and tragedies of its own, surely we have only to consider the implications of an Axis victory to recognise the magnitude of the Allied triumph.'[14] New Zealand contributed as much as it possibly could towards an Allied victory and took considerable risks in doing so. It was, as Michael King put it, the 'last great common denominator, the last intense experience that tens of thousands of people would share, and one whose rationale was accepted by the country as a whole'.[15]

This photographic record reveals much about a critical period in New Zealand's history when most of the country was united in a common cause while, for six long years, all hell really was let loose.

ABOVE | Air-to-air view of a No. 98 Squadron Mitchell bomber heading off on a mission in 1944.

AIR FORCE MUSEUM OF NEW ZEALAND, ALB010711B283

ABOVE | Hitler causes a war. German troops advance into Poland in September 1939. This photograph and the three that follow are from a German propaganda album held at the National Army Museum in Waiouru.

NATIONAL ARMY MUSEUM, 2000-973

| CHAPTER 1 |

At War Again

On 1 September 1939, after fabricating an attack by Polish soldiers on a German radio station, German military forces invaded Poland. This deliberate move, designated 'Case White' by Germany, was the culmination of the diplomatic and international crisis that had gripped Europe since the Nazis and Adolf Hitler came to power in January 1933. Determined to reverse the outcome of the First World War and to dominate Europe, Nazi Germany had set out on a path that was certain to lead to war. After allowing Germany generous concessions under their policy of appeasement, which aimed to prevent war by redressing some of the grievances Germany felt about the Treaty of Versailles that ended the First World War — including the remilitarisation of the Rhineland, the right to rearm, the forced union with Austria and the occupation of Czechoslovakia — Britain and France had finally decided enough was enough. Poland would be their line in the sand. On 31 March 1939, both nations announced a unilateral guarantee of support to Poland should it be attacked. Similar guarantees were also later offered to Greece and Romania.

Hitler, however, was determined to have his war and to take the Lebensraum, the territory in Eastern Europe to which he felt the Germans were entitled. On 22 August he announced that Germany and the Soviet Union had signed a non-aggression pact. This pact between two obvious enemies was 'a diplomatic coup . . . that shocked the world'.[1] A secret protocol of the pact divided Eastern Europe, including Poland, between the two countries. Hitler now felt that the way was clear for Germany's invasion of Poland. He did not expect Britain and France to honour their commitment; it was a critical miscalculation.

The British government, determined to stand firm on this latest violation of international law, immediately issued an ultimatum for Germany to begin withdrawing its forces. When this expired at 11 a.m. on 3 September 1939 a disillusioned and gloomy Neville Chamberlain announced to the world that Britain was at war with Germany.

Britain's decision left the New Zealand government with a grave choice. Chamberlain's declaration of war made it clear that he was speaking for the British government alone. He announced in the House of Commons that 'this country' and, not the Commonwealth, was at war with Germany.[2] The dominions, which included New Zealand, were regarded as autonomous communities responsible for their own domestic and external affairs. In September 1939 they faced a stark choice: either join Britain at war or declare their neutrality.

Neither New Zealand nor Australia hesitated. Australia alone of the dominions, under its Anglophile Prime Minister Robert Menzies, adhered to the principle that Britain's declaration of war applied to all the king's loyal subjects. This meant that Australia was automatically at war. It was surprising that Menzies, trained in the law, made such a constitutional error. New Zealand's response, though it followed parliamentary procedure, was equally speedy. The Cabinet met as soon as the formal notification from Britain was received, just before midnight on 3 September. The next day the governor-general dispatched a cable to London, stating that New Zealand was also at war with Germany and had been from the time the ultimatum expired. New Zealand was one of the first democratic nations to enter the war, alongside Britain, France and Australia. South Africa joined the war effort three days later and Canada did so after seven days. Éire was the sole British dominion that remained neutral throughout the war.

IT WOULD BE A MISTAKE to conclude that New Zealand's rapid declaration implied little thought was given to such a grave matter or, even worse, that it was a case of blind obedience. New Zealand was very much part of the British world and it was in New Zealand's interests to protect it. Historian Ian McGibbon has identified three key influences behind the government's decision and why it was almost universally supported: a desire to stand with British kith and kin; a need to protect New Zealand's economic and physical security, which was heavily dependent on Britain; and the recognition that aggression by lawless states had to be resisted. It was a combination of self-interest and ideology. 'Each on its own might have impelled New Zealand into the war; together they amounted to a powerful basis for action, and ensured almost universal acceptance of the government's decision for war.'[3] These reasons, and New Zealand's independent stance, were stressed in one of the most impressive and important speeches ever made by a New Zealand prime minister. On 5 September 1939, an ailing Michael Joseph Savage said:

> *Not in anger but in sorrow, not in light heartedness, but with heavy hearts, not in hatred but with a grave sense of great responsibility to mankind and to the future of humanity, not in malice and revenge,*

ABOVE | The cavalry rides out of shot as Hitler takes the salute. Despite the propaganda, the German Army in 1939 was primarily an infantry force and still depended on horses for its mobility.

NATIONAL ARMY MUSEUM, 2000-973

ABOVE | German soldiers goose-step in a victory parade, 1939.

NATIONAL ARMY MUSEUM, 2000-973

RIGHT | German Mark II light tanks on parade in Warsaw, Poland, in 1939; Hitler takes the salute in the centre background.

NATIONAL ARMY MUSEUM, 2000-973

ABOVE | Departing to fight another war: New Zealand soldiers wave goodbye as their troopship pulls away from the Wellington wharves in 1940.

NATIONAL ARMY MUSEUM, 1987-1828

LEFT | Using every possible vantage point for a last look at family and homeland.

NATIONAL ARMY MUSEUM, 1987-1828

but with a prayer of peace on our lips, the British people today dedicate themselves to the work of overthrowing the oppressor and freeing the peoples of the earth from bondage and slavery to a ruthless and cruel tyranny . . . Both with gratitude for the past, and with confidence in the future, we range ourselves without fear beside Britain. Where she goes, we go, where she stands, we stand. We are only a small and young nation, but we are one and all a band of brothers, and we march forward with a union of hearts and wills to a common destiny.[4]

Unlike 1914, there was no widespread enthusiasm for another war. There were no cheering crowds or patriotic songs. New Zealanders were no longer innocent; the First World War had shown them that fighting a country as powerful as Germany meant years of pain, separation and suffering. It meant appalling numbers of dead and wounded men and lasting physical and mental pain for those who survived. It meant the destruction of families.

THERE WAS ANOTHER reason for the sombre mood of September 1939, although few at the time admitted to it: 'New Zealand for the first time faced a struggle in which the outcome was obviously distant and even uncertain.'[5] In contrast to 1914, there was considerable doubt whether Britain, its dominion allies and France could prevail over Nazi Germany. The British and French military forces were nowhere near as strong as they had been before the Great War. Added to this was the likelihood that Italy, which had recently concluded the Pact of Steel with Germany, and Japan, might join the enemy. The Soviet Union was also an unknown quantity. It was either now a German ally or would acquiesce in Germany's conquests. The remaining great power, the United States, was committed to remaining neutral. The international and military situation in 1939 did not look good.

ABOVE | Members of 27 Machine Gun Battalion ready to embark from Lyttelton with the First Echelon in 1940.

STEPHEN PARSONS COLLECTION

RIGHT | The *Sobieski*, pictured here, and the *Dunera* left Lyttelton on 5 January 1940, carrying the South Island members of the First Echelon. The North Island troops departed from Wellington on the same day on the *Empress of Canada*, *Strathaird*, *Orion* and *Rangitata*. All the transports sailed together to Egypt.

NATIONAL MUSEUM OF THE ROYAL NEW ZEALAND NAVY, OC 624

ABOVE | In September 1939, New Zealand did not have its own navy, but was a flotilla of the Royal Navy numbering fewer than 1000 men. The Royal New Zealand Navy, established in October 1941, had more than 70 vessels in 1945. Here, newly recruited sailors receive instruction in boat drill at the training depot HMNZS *Philomel* in Devonport, Auckland.

NATIONAL MUSEUM OF THE ROYAL NEW ZEALAND NAVY, OC 285

| CHAPTER 2 |

Early Days

Declaring war in September 1939 had been swift and relatively straightforward. Enlisting enough men to fight in a long, total war, and then sustaining those forces over the conflict's duration, caused the New Zealand government considerable problems. The situation was not helped by the fact that the country was not as well prepared for war in 1939 as it had been in 1914.

Of the three services, the navy was the best placed to make an immediate contribution to the war effort and soon did so in 'spectacular fashion' at the Battle of the River Plate in December 1939. But the navy, technically the New Zealand Division of the Royal Navy until 1 October 1941, when it became the Royal New Zealand Navy, was small, and it lacked sufficient trained New Zealand personnel. The crew of its main strike force, the two modern light cruisers, *Achilles* and *Leander*, were supplemented by men on loan from the Royal Navy — in the case of *Achilles* almost half the crew and its captain.[1] In 1939, only eight of the 82 officers were New Zealanders; of the 1257 ratings, 541 were British personnel. With more vessels needed to protect New Zealand waters, and Britain also in dire need of trained sailors, from September 1939 New Zealand needed to recruit and train as many seamen as possible. By the end of the war some 10,000 men and 519 women had served with New Zealand naval forces.[2]

By the time war broke out, the Royal New Zealand Air Force (RNZAF), established in 1937, had undergone considerable expansion. The plan, in September 1938, was to train 1000 pilots for the RAF, but in May 1939 this was changed: instead the RNZAF would provide approximately 650 pilots, 300 observers and 350 air-gunners.[3] On the outbreak of war 550 New Zealanders were serving in the RAF, the most of any dominion. This accounts for the large number of New Zealanders who fought in the 1940 Battle of Britain, the biggest contingent of non-Britons after Polish pilots. New Zealand also waived its purchase of the first six Wellington bombers for its own use; these instead became, along with their crews, the nucleus of No. 75 (Bomber) Squadron of the RAF. Through the Empire Air Training Scheme (EATS), and through direct entry, some 10,950 New Zealanders served in the RAF in nearly every theatre of war.[4]

A further 45,000 served in the RNZAF in New Zealand and in the Pacific.[5]

On 13 September 1939, just 10 days after declaring war, the New Zealand government offered to provide Britain with a 'fully trained division' to be available for service in any theatre of war 'within a period of eight months from today'. It did, however, note that Britain would need to provide most of the division's war equipment and arrange 'safe and adequate means of transport'.[6] Even with these stipulations this offer was unrealistic. The army was 'the Cinderella of the armed services' and, according to Kippenberger, 'all but went out of existence' during the 1930s.[7] It would need to be built up from scratch in order to provide an expeditionary force that could make a reasonable contribution to the war effort. Fortunately, though, when recruiting for an expeditionary force began on 12 September 1939, volunteers flocked to join up — almost 60,000 of them before conscription came into force on 22 July 1940. By the end of that first day of recruiting some 5419 had enlisted.[8]

The government, lacking faith in the senior commanders in New Zealand, appointed Bernard Freyberg as the commander of 2 NZEF and the infantry division that would be its main fighting component. Freyberg had offered his services in September 1939, stating that he 'would be glad to serve with compatriots again'.[9] This was somewhat ironic, as Freyberg had not previously been on active service with New Zealand forces. Although he had a distinguished war record, including the award of the Victoria Cross on the Somme in 1916, he was very much a British officer, having served with the Royal Naval Division and the Grenadier Guards.[10]

THE FIRST ECHELON of the division, which would later become 4 Brigade, entered camp on 3 October and departed for Egypt on 5 January 1940, arriving there six weeks later. The Second Echelon, which became 5 Brigade, left New Zealand on 2 May 1940. As it sailed the war situation deteriorated — France had been invaded and was facing defeat, and Italy would soon be an enemy combatant — and so the Second Echelon was diverted to a vulnerable and isolated Britain. The New Zealanders' arrival on 16 June 1940, together with a force of Australians, was a considerable morale booster, and was promoted as an example of Commonwealth unity. Even Freyberg, who had not wanted the Second Echelon diverted from Egypt, admitted that the change of plan 'had been most opportune and had steadied the nation considerably'.[11] The Third Echelon (6 Brigade) left New Zealand on 27 August 1940 and reached

RIGHT | Petty Officer 'Iggie' Biggs instructs a class as part of their wireless telegraphy training at HMNZS *Philomel*. The sailors receiving instruction are, from left, unknown, Gordon Woods, Ron Campbell, Brian Taylor, Crichton, with Tony Baxendale in the front.

NATIONAL MUSEUM OF THE ROYAL NEW ZEALAND NAVY, OC 287

LEFT | The Minister of Defence, Frederick Jones (third from left), and Acting Prime Minister Walter Nash (fourth from left), pose with four direct-entry naval cadets: from left, Maxson McDowell, William Petersen, John Armstrong and Edward Thorne.

NATIONAL MUSEUM OF THE ROYAL NEW ZEALAND NAVY, OC 371

Egypt two months later. It was not until 12 January 1941 that 5 Brigade finally sailed from Britain to Egypt, which meant New Zealand finally had a complete, fully equipped but partially trained infantry division in an active theatre of operations. On 23 February 1941, 18 months after the division's inception, Freyberg informed the government that two brigades were 'now fit for war' and that 5 Brigade, after being refitted and 'hardened up after the voyage', would soon reach this state. Freyberg concluded his cable: 'Therefore, I feel that should the British Government request the release of the New Zealand Expeditionary Force for a full operational role the New Zealand Government can now do so with confidence.'[12] As W. David McIntyre has noted, 'That it took so long was a comment on the lack of a trained professional army in peacetime and the shortage of training facilities in New Zealand. This was the most striking wartime effect of defence policy between the wars.'[13]

The Second New Zealand Expeditionary Force and its fighting component, which in June 1942 became 2 New Zealand Division, was the largest manpower contribution of the country's war effort. All told, 104,000 New Zealanders served with 2 NZEF during the war.[14] Once conscription began in mid-1940, the thousands who had volunteered to serve in army khaki were joined, over the next five years, by another 306,000 men. Most joined 2 NZEF and later 3 NZEF, which served in the Pacific, or the land forces retained in New Zealand. Almost 50,000 opted for the RAF or RNZAF, and a much smaller number, just over 10,000, signed up with New Zealand naval forces or the Royal Navy for a sea combat role.

ABOVE | In April 1940 the New Zealand High Commissioner, William (Bill) Jordan, inspects the first draft of New Zealanders for the Royal Naval Volunteer Reserve.

NATIONAL MUSEUM OF THE ROYAL NEW ZEALAND NAVY, OC 192

LEFT | From mid-1940 a steady stream of New Zealanders arrived in Britain to begin their navy training. These volunteers pose in front of a statue of Lord Nelson at HMS *Ganges*.

NATIONAL MUSEUM OF THE ROYAL NEW ZEALAND NAVY, OC 179

ABOVE | Able Seaman Jack True, enjoying Trafalgar Square in 1940, was killed when the cruiser HMS *Neptune* was sunk in the Mediterranean in December 1941.

NATIONAL MUSEUM OF THE ROYAL NEW ZEALAND NAVY, 2006.186.16. P7.1

ABOVE | New Zealanders who were part of the Windsor Guard perform a haka outside a restaurant to impress the waitresses. It must have worked, as they were given a cup of tea for their efforts.

NATIONAL MUSEUM OF THE ROYAL NEW ZEALAND NAVY, 2014.32.23.P9.1

RIGHT | The 'NZ boys rendezvous' at a London pub. They are crewmates of Jack Delaney, on the far left, who served in No. 851 Naval Air Squadron of the Fleet Air Arm.

NATIONAL MUSEUM OF THE ROYAL NEW ZEALAND NAVY, 2014.32.23.P16.1

RIGHT | A near miss on a merchant vessel in the English Channel in July 1940. The Merchant Navy is the only civilian organisation honoured in the New Zealand National War Memorial in Wellington. About 3000 New Zealanders served in the Merchant Navy during the war.

NATIONAL MUSEUM OF THE ROYAL NEW ZEALAND NAVY, OC 1042

BELOW | *Achilles* at Port Fitzroy, Great Barrier Island, in 1939, just months before the light cruiser was involved in the first New Zealand military action of the war: the Battle of the River Plate in December 1939.

NATIONAL MUSEUM OF THE ROYAL NEW ZEALAND NAVY, OC 1162

ABOVE | *Achilles*' opponent during the Battle of the River Plate was the *Admiral Graf Spee*, a Panzerschiff (armoured ship) of the Deutschland class, known to the Allies as a pocket battleship. Along with the heavy cruiser *Exeter* and the light cruiser *Ajax*, *Achilles* made up the South American Division of the Royal Navy, but the trio was outclassed by the firepower *Graf Spee* possessed.

NATIONAL MUSEUM OF THE ROYAL NEW ZEALAND NAVY, ACN 0151

LEFT | Hunting for the *Graf Spee*. HMS *Ajax* photographed from on board HMNZS *Achilles*.

NATIONAL MUSEUM OF THE ROYAL NEW ZEALAND NAVY, AUG 0188

ABOVE AND BELOW | The 6-inch guns of *Achilles*' X and Y turrets in action during the Battle of the River Plate.

NATIONAL MUSEUM OF THE ROYAL NEW ZEALAND NAVY, AUG 092, AUG 0195

ABOVE | The control tower of *Achilles*, showing shrapnel damage from the *Graf Spee*'s guns.

NATIONAL MUSEUM OF THE ROYAL NEW ZEALAND NAVY, AUG 0219

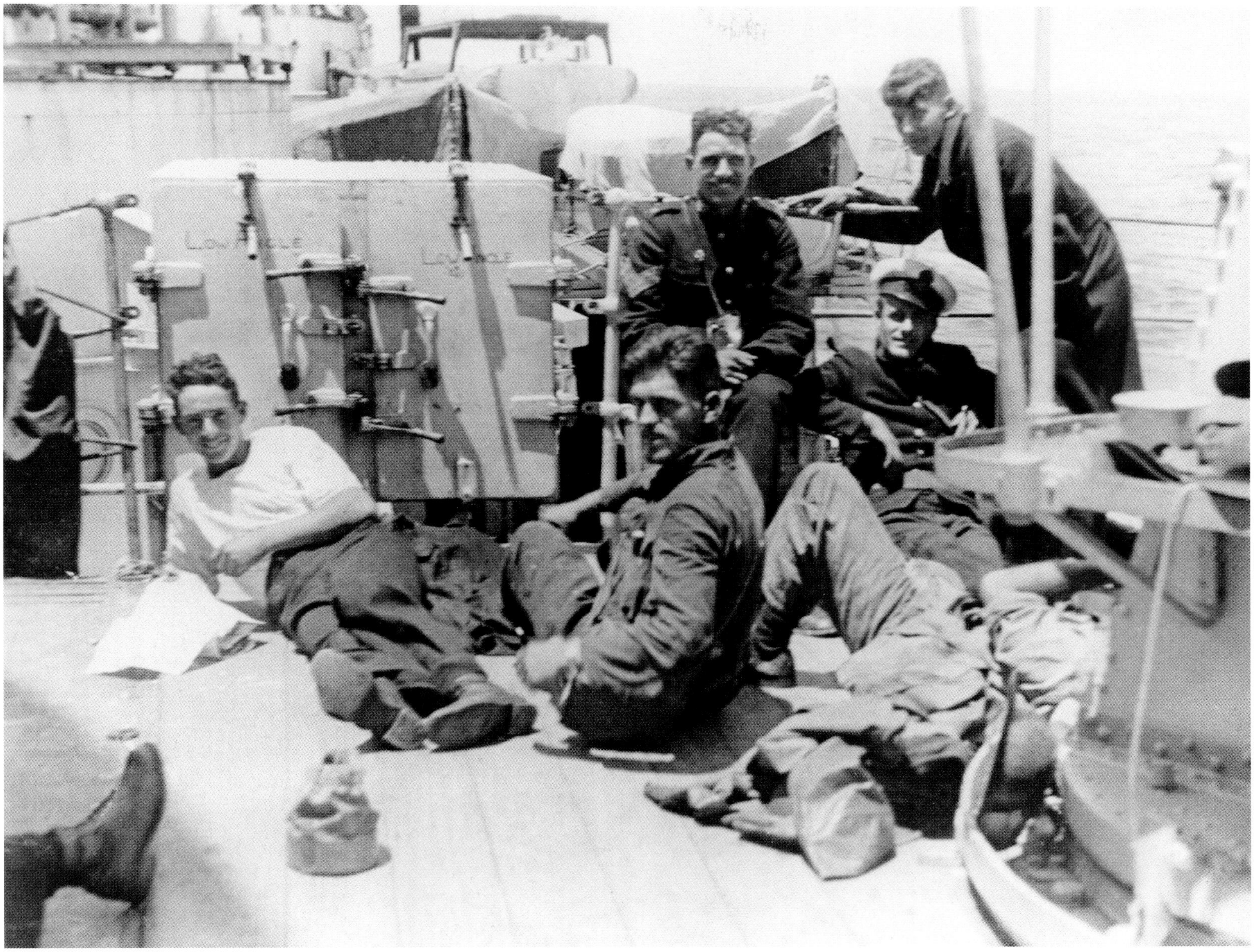

ABOVE | *Achilles* crew relax during a lull in the battle. They are, from left, Albert Chapman, Milton Hill, Sergeant R. M., T. Maguire and Ron Batkin.

NATIONAL MUSEUM OF THE ROYAL NEW ZEALAND NAVY, AUG 0234

RIGHT | A lightly damaged *Graf Spee* in Montevideo Harbour, Uruguay. Under international law, the ship could not remain long in this safe haven.

NATIONAL MUSEUM OF THE ROYAL NEW ZEALAND NAVY, AUG 0006

ABOVE | Trapped in the harbour and believing that a much larger naval force had been assembled to destroy his ship, the German captain, Hans Langsdorff, ordered his ship *Graf Spee* to be scuttled. Langsdorff committed suicide three days later.

NATIONAL MUSEUM OF THE ROYAL NEW ZEALAND NAVY, AUG 0133

RIGHT | *Achilles* crew take a last look at the *Graf Spee* before departing for home.

NATIONAL MUSEUM OF THE ROYAL NEW ZEALAND NAVY, AUG 0251

LEFT | Before returning to New Zealand, *Achilles* transfers its wounded ashore at Port Stanley in the Falkland Islands.

NATIONAL MUSEUM OF THE ROYAL NEW ZEALAND NAVY, AUG 0246

LEFT | A victorious *Achilles* arrives in Auckland on 23 February 1940.

NATIONAL MUSEUM OF THE ROYAL NEW ZEALAND NAVY, AAF 0121

ABOVE | The *Leander* docked at Wellington in May 1940. It sailed on 2 May 1940 as an escort for the Second Echelon troopships.

NATIONAL MUSEUM OF THE ROYAL NEW ZEALAND NAVY, 2006.186.142. P3

ABOVE | HMNZS *Leander* experiencing a rough Tasman crossing.

NATIONAL MUSEUM OF THE ROYAL NEW ZEALAND NAVY, 2006.186.147. P5.1

RIGHT | A dramatic photograph taken from the deck of RMS *Rangitata* in June 1940 shows the tanker attached to the convoy has been hit by a torpedo.

AIR FORCE MUSEUM OF NEW ZEALAND, ALB0224349040

RIGHT | The first stage of turning civilians into soldiers: a group of volunteers leaves for training camp.

NATIONAL ARMY MUSEUM, 2002-566

ABOVE | Men recruited for war service were often at their peak of physical fitness. Here Mich Sullivan does his 'strong man' act.

O. E. B. (OZ) TURLEY COLLECTION

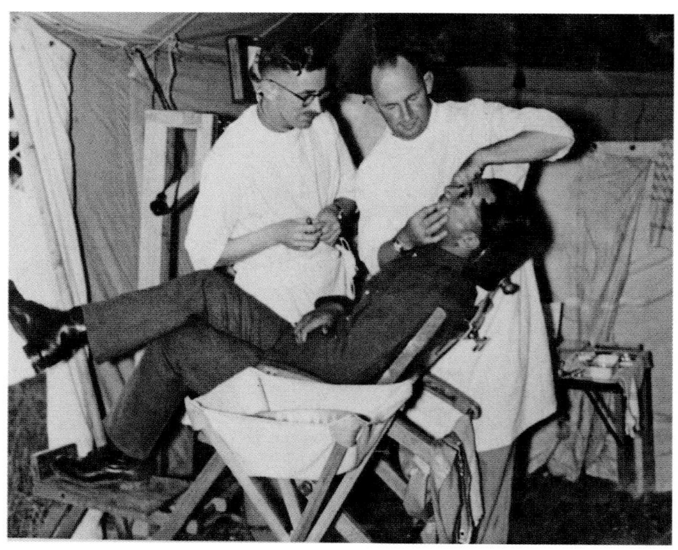

RIGHT | Being physically fit included healthy teeth. A new recruit undergoes dental treatment at Palmerston North Camp.

NATIONAL ARMY MUSEUM, 2001-1088-1

ABOVE | Members of 27 Machine Gun Battalion enjoy the fine weather at Cave, South Canterbury, in 1939. Note the machine gun and ammunition boxes. The officer in the centre wearing the Sam Browne belt is 2nd Lieutenant David Parsons.

STEPHEN PARSONS COLLECTION

LEFT | New soldiers enjoying army life on a training exercise at Cave, South Canterbury, in 1939.

NATIONAL ARMY MUSEUM, 2006-49

ABOVE | Those training at Waiouru take a light-hearted view of flooding in the camp.

NATIONAL ARMY MUSEUM, 1993-1193

LEFT | Waiouru Military Camp, a harsh, desolate place near Mount Ruapehu, was established in 1940 to train soldiers and toughen them up.

NATIONAL ARMY MUSEUM, 1993-1329

LEFT | Soldiers train on the 6-inch gun at Wellington's Fort Dorset in 1939.

NATIONAL ARMY MUSEUM, 2001-720

ABOVE | Learning to use bayonets at a New Zealand military camp.

NATIONAL ARMY MUSEUM, 2018-160-4

RIGHT | Machine gun training at Trentham Military Camp in Upper Hutt.

NATIONAL ARMY MUSEUM, 2013-143-1

RIGHT | On a chilly day, a crowd watches as the salute is taken in a military parade outside the Wellington Railway Station.

NATIONAL ARMY MUSEUM, 1989-393

BELOW | Soldiers parade down Wellington's Lambton Quay before departing for overseas.

NATIONAL ARMY MUSEUM, 1989-431

BELOW | On their last day in Trentham Camp, these men look both relieved and tired.

NATIONAL ARMY MUSEUM, 2002-566

RIGHT | New Zealand soldiers on a troopship feel the tropical heat.

NATIONAL ARMY MUSEUM, 1990-389-2

BELOW | Two-up, the traditional Australian gambling game popular with Anzac troops in the First World War, helped to pass the time on board.

NATIONAL ARMY MUSEUM, 1990-389-1

ABOVE | The First Echelon lined up in front of Parliament House on 3 January 1940.

NATIONAL ARMY MUSEUM, 1992-164

BELOW | It was not only men who sailed away to war. Betty Lidgett embarked on HMS *Dunnottar Castle* as a private in the Women's Auxiliary Army Corps.

NATIONAL ARMY MUSEUM, 2010-287-2

RIGHT | Muriel Dodd from Mititai in Northland also served overseas as a private in the Women's Auxiliary Army Corps.

NATIONAL ARMY MUSEUM, 1997-172-1

RIGHT | Gunner Ivan Collins strikes a match to light Anthony Eden's cigarette on board the troopship *Empress of Canada*, which had left New Zealand on 5 January, carrying the First Echelon. When it arrived at Port Said on 12 February 1940, Eden, as secretary of state for dominion affairs, was there to welcome the men. This photograph later appeared in a New Zealand news magazine as one of 'The Month's Best Anzac Pictures' with the caption 'Match, Tony'. Eden gave Collins a cigarette in return.

COLLINS COLLECTION

LEFT | London-born, Bernard Freyberg came to New Zealand at the age of two and grew up there, but served with the British forces in the First World War. In the Second World War he commanded the Second New Zealand Expeditionary Force and 2 New Zealand Division.

NATIONAL ARMY MUSEUM, 1993-1313-2

RIGHT | General and Lady Freyberg. During the war Barbara Freyberg ran the New Zealand Forces Club established in Cairo by her husband. Serving tea and refreshments, it was staffed by young women known as Tuis, who were part of the Women's Auxiliary Army Corps.

NATIONAL ARMY MUSEUM, 2010-287-2

BELOW | New Zealand troops are barged ashore in Suez Harbour. Egypt would be their main training area for the next two years.

NATIONAL ARMY MUSEUM, 1990-389-1

RIGHT | One attraction of volunteering for military service was a chance to see the world. New Zealand nurses and soldiers enjoy a day of sightseeing in Egypt, probably at the Mortuary Temple of Hatshepsut, Deir el-Bahari. Eighteen nurses sailed with the First Echelon.

NATIONAL ARMY MUSEUM, 1992-712

ABOVE | With the Great Sphinx of Giza and a pyramid behind them, New Zealand soldiers smile for the camera.

NATIONAL ARMY MUSEUM, 1993-1329

RIGHT | Two nurses pose proudly on top of a pyramid.

NATIONAL ARMY MUSEUM, 1993-2420-1

ABOVE | Egypt was not just a place to see the sights. It was where the New Zealand soldiers began their hard training for the campaigns ahead. Here they learn the workings of the Bren gun.

ENGINEER CORPS MEMORIAL CENTRE, WALLACE 190

ABOVE | New Zealand machine gunners carry their equipment during a field exercise.

NATIONAL ARMY MUSEUM, 2016-339-1

ABOVE | A soldier stops to watch ships pass through the Suez Canal at Kantara, Egypt.

ENGINEERS CORPS MEMORIAL CENTRE, 9768 096

RIGHT | Three New Zealand nurses make the most of their time in Egypt, with the help of a local Egyptian guide.

NATIONAL ARMY MUSEUM, 1992-712

RIGHT | Anti-aircraft training with a Bren gun at Helwan Camp, Egypt.

NATIONAL ARMY MUSEUM, 1993-1313-1

LEFT | Soldiers enjoy their lunch break at the Helwan Camp rifle range.

NATIONAL ARMY MUSEUM, 2016-339-1

RIGHT | Members of the First Echelon listen to the BBC news in Maadi Camp, Egypt, in 1940.

COLLINS COLLECTION

RIGHT | While the First and Third Echelons sweated it out in Egypt, soldiers of the Second Echelon (5 Brigade) had a more relaxed and pleasant time in Britain. These soldiers are enjoying their first mail from New Zealand.

NATIONAL ARMY MUSEUM, 2009-1052

ABOVE | A 'well earned rest after manoeuvres' says the caption, but the soldiers do not look too stressed.

NATIONAL ARMY MUSEUM, 2009-1052

RIGHT | Three soldiers of 5 Brigade gather plums in the mild Kent summer.

NATIONAL ARMY MUSEUM, 2009-1052

ABOVE | The 5 Brigade Band assembles at the double for a practice.

NATIONAL ARMY MUSEUM, 2002-566

RIGHT | Meanwhile, in Egypt, New Zealand troops on a route march near Maadi Camp.

NATIONAL ARMY MUSEUM, 1990-1010

ABOVE | Digging in in the desert. This flat landscape offered very little cover.

NATIONAL ARMY MUSEUM, 1990-1010

BELOW | Trucks queuing for petrol.

NATIONAL ARMY MUSEUM, 1990-1010

RIGHT | Members of the New Zealand Railway Company dig out sand from the switch points. Trains were essential for bringing supplies and water, and an easy target for the enemy. Often forgotten, Kiwi railwaymen played a vital role in the desert campaign.

NATIONAL ARMY MUSEUM, 1990-389-3

ABOVE | The men of 27 Battery of 5 Field Regiment calibrate their guns.

NATIONAL ARMY MUSEUM, 2005-180

LEFT | Having a bath in the desert.

NATIONAL ARMY MUSEUM, 1990-1010

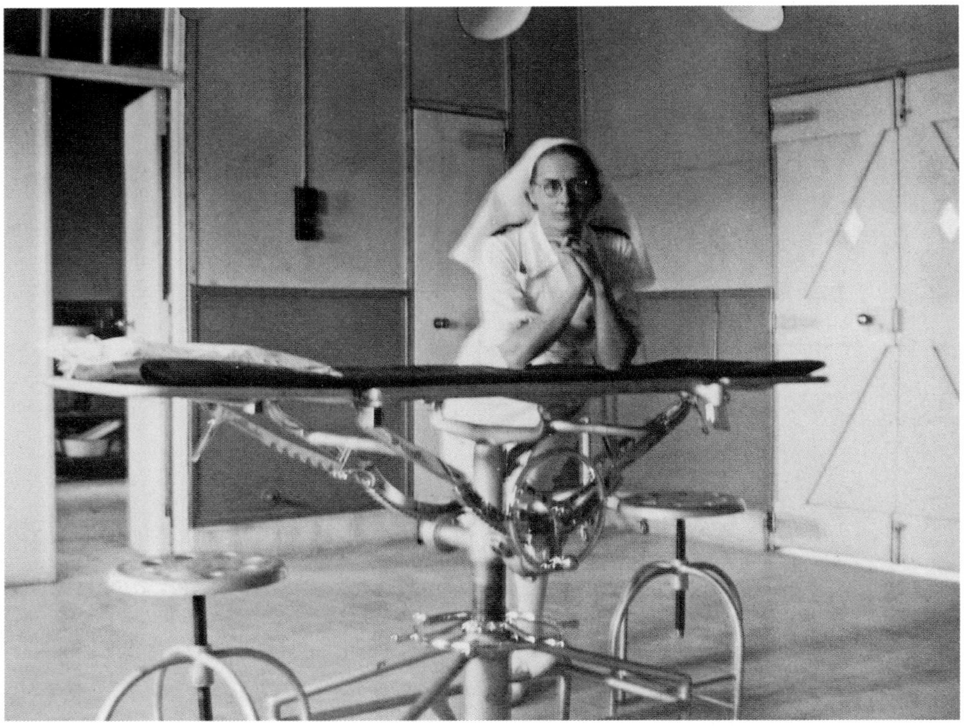

LEFT | Staff Nurse Violet Wallace from Stratford in the operating theatre of a New Zealand hospital in Egypt.

NATIONAL ARMY MUSEUM, 1993-2420-1

BELOW | Now British foreign secretary and future prime minister, Anthony Eden pays a visit to No. 2 New Zealand General Hospital in Cairo. On the left, escorting Eden, is First World War veteran Colonel Frederick Montgomery Spencer, who died of typhus on 12 June 1943.

NATIONAL ARMY MUSEUM, 1993-2420-1

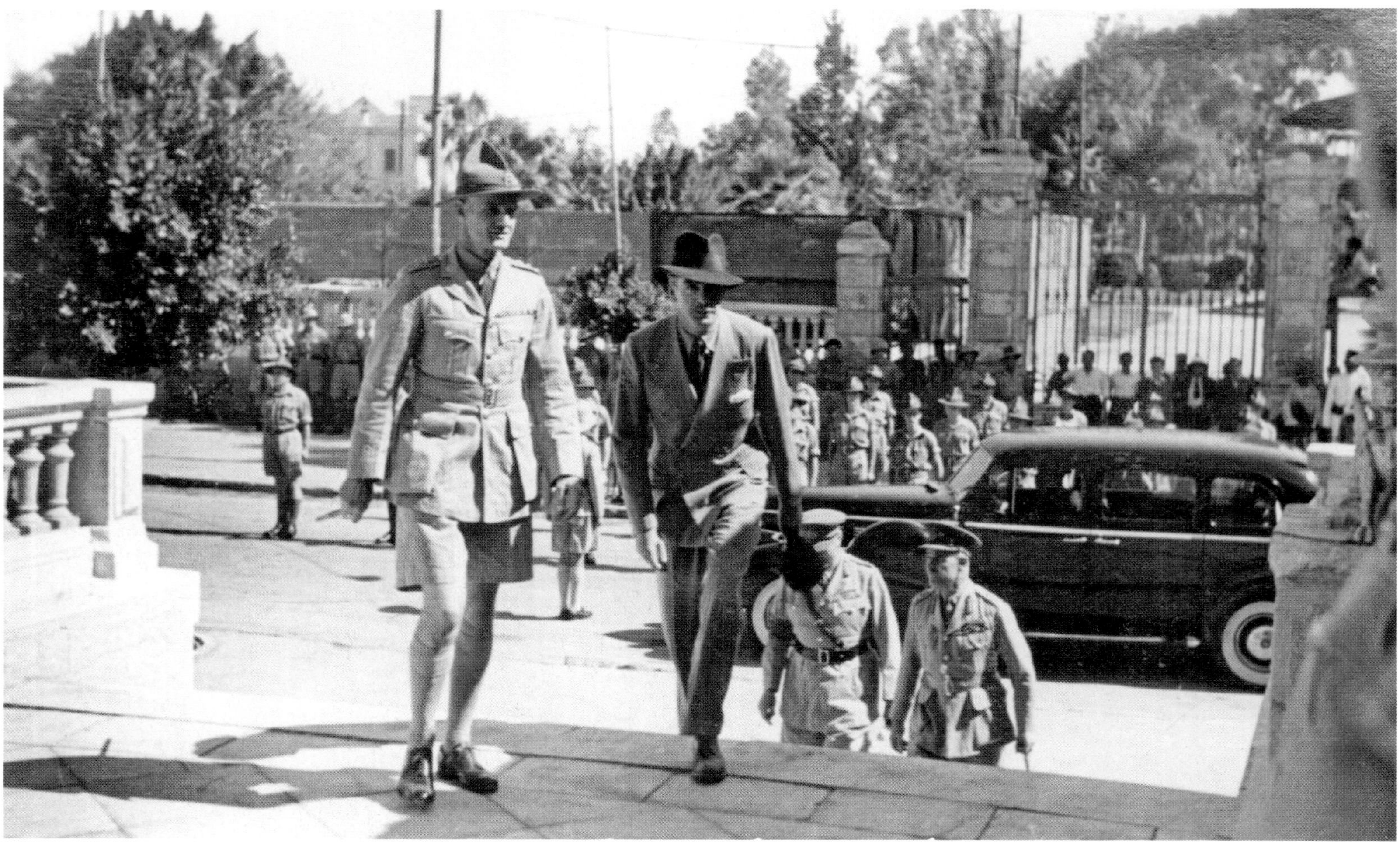

ABOVE | New Zealand nurses of No. 2 New Zealand General Hospital waiting for Anthony Eden's visit.

NATIONAL ARMY MUSEUM, 1993-2420-1

RIGHT | Famous actor Vivien Leigh and members of the Old Vic Spring Party that toured North Africa in 1943 visit a New Zealand field hospital. The man on the left is most likely Hugh 'Binkie' Beaumont, a theatre manager and producer who organised the tour.

NATIONAL ARMY MUSEUM, 1992-712

LEFT | Soldiers washing clothes in the desert. Keeping body and clothes clean, and managing to shave, were major challenges with water at such a premium.

NATIONAL ARMY MUSEUM, 1990-389-3

BELOW | Digging duty at Maadi Camp.

NATIONAL ARMY MUSEUM, 1993-1315

RIGHT | New Zealand soldiers put their bayonets to use in a mock battle at Jaraburg.

NATIONAL ARMY MUSEUM, 1993-1352-1

ABOVE | In contrast to the searing daytime heat, the desert was freezing cold at night. In their balaclavas and greatcoats, this trio warms up with breakfast and a hot cup of tea.

NATIONAL ARMY MUSEUM, 1992-1084-7

LEFT | Wagh El Birket, the notorious red-light district of Cairo, was known to the troops as the Berka (or Birka). Most New Zealand soldiers went to have a look, even if they did not set foot in a brothel. The British military authorities closed the area down in May 1942.

NATIONAL ARMY MUSEUM, 1992-164

LEFT | As they had at Gallipoli, flies made the men's lives a misery, settling on their food and landing on their lips, seeking moisture. These veils offered some protection.

NATIONAL ARMY MUSEUM, 2001-720

BELOW | High Commissioner Bill Jordan, teacup in hand, enjoys a visit with soldiers in Maadi Camp. Throughout the war he worked tirelessly for New Zealand service personnel.

NATIONAL ARMY MUSEUM, 2008-496

ABOVE | Three nursing sisters, with brimmed hats, and another uniformed woman enjoy the obligatory camel ride at Mena in 1940. The medical staff are named as Sisters Bolton and Crawford and Matron Eva Mackay, who was in charge of No. 1 New Zealand General Hospital.

NATIONAL ARMY MUSEUM, 2008-496

LEFT | Tea, beer and buns at the opening of Lowry Hut at Maadi, named after wealthy Hastings racehorse breeder Thomas Lowry, who donated a large amount of money for its construction. The only building in the camp designed for desert conditions, it was much bigger than a hut: it had a comfortable lounge, a stage and rooms for various recreational activities. Lowry's daughter, Gertrude, known as Beet, worked there throughout the war.

NATIONAL ARMY MUSEUM, 2008-496

ABOVE | New Zealand Engineers, life jackets in place, practise bridge building in Egypt.

ENGINEER CORPS MEMORIAL CENTRE, RNZE HISTORY 29

RIGHT | Sapper Thomas Bane competes in the hop, skip and jump event at a sports day in Cairo.

ENGINEER CORPS MEMORIAL CENTRE, 18 ATC

LEFT | Major-General Bernard Freyberg, a champion swimmer, dives in for the first length of a swimming pool built at Maadi by the New Zealand Engineers. The pool, which was Freyberg's idea, was opened on 5 April 1940.

ENGINEER CORPS MEMORIAL CENTRE, 9527 023

ABOVE | The staff of No. 2 New Zealand General Hospital and guests enjoy a wedding breakfast in Cairo to celebrate the marriage of one of their nurses.

NATIONAL ARMY MUSEUM, 1993-2420-2

RIGHT | Confetti in Cairo — quite a few nurses married soldiers while serving overseas.

NATIONAL ARMY MUSEUM, 1992-712

RIGHT | The annotations show what happened to these seven officers. Only one did not become a casualty or prisoner of war.

NATIONAL ARMY MUSEUM, 1993-1313-1

BELOW | Lieutenant-Colonel Howard Kippenberger, fourth from left in the front row, poses in Cairo with the officers of 20 Infantry Battalion. Of the men in this photograph, 11 were killed and 11 wounded, including Kippenberger; eight became prisoners of war.

NATIONAL ARMY MUSEUM, 2015-40-19

ABOVE | New Zealand soldiers on leave with a guide to show them the sights. This photograph was taken on 4 November 1940.

STEPHEN PARSONS COLLECTION

BELOW | David Parsons sent festive greetings to his father in New Zealand in 1940. He spent his first desert Christmas at Baggush.

STEPHEN PARSONS COLLECTION

ABOVE | Lieutenant David Parsons with local children.

STEPHEN PARSONS COLLECTION

ABOVE | An air-to-air view of two Fairey Gordons flying in formation over the Port Hills and Christchurch. Most New Zealand pilots learned to fly in biplanes like Fairey Gordons, Baffins and Tiger Moths.

AIR FORCE MUSEUM OF NEW ZEALAND, ALB922607022

RIGHT | A formation of Baffins over the Wairau Valley near Blenheim.

AIR FORCE MUSEUM OF NEW ZEALAND, ALB952622042

RIGHT | Flying Officer William Christiansen about to climb into a Tiger Moth at RNZAF Station Tairei, in 1940. Christiansen, who later served with No. 67 Squadron RAF in the air battles over Burma, was reported missing, believed killed, on 9 April 1943. He was 22 years old.

AIR FORCE MUSEUM OF NEW ZEALAND, ALB902765A001

RIGHT | Oxford trainers high above the Canterbury Plains.

AIR FORCE MUSEUM OF NEW ZEALAND, ALB841655079

BELOW | A flight of airmen, rifles shouldered, at their passing-out parade at Wigram in June 1939.

AIR FORCE MUSEUM OF NEW ZEALAND, MUS9700410

RIGHT | Learning to fly was a dangerous activity. During the First World War more British pilots were killed in accidents than from combat. Training and other flying accidents were common in the Second World War and often caused fatalities. The two occupants of Fairey Gordon NZ620, Pilot Officer George Brabyn and Leading Aircraftman J. G. Hannah, survived when it crashed just north of the Wigram airfield in April 1940. Many others were not so fortunate.

AIR FORCE MUSEUM OF NEW ZEALAND, ALB922607024

LEFT | Tiger Moth NZ749 from No. 2 Elementary Flying School at New Plymouth lies on its back after a disastrous landing in June 1940. The pilot, Leading Aircraftman Joseph Parry from Paeroa, survived this crash but was killed on 29 September 1941 while flying with No. 99 Squadron RAF.

AIR FORCE MUSEUM OF NEW ZEALAND, ALB8485871B019

ABOVE | 'Some of the New Zealand boys in our course', reads the original annotation on this photograph. The course was at the No. 10 Flying Training School at Dauphin in Manitoba in 1942. The New Zealanders were part of the Empire Air Training Scheme (EATS), most of which was based in Canada.

AIR FORCE MUSEUM OF NEW ZEALAND, ALB031891055

RIGHT | Canadian Minister for National Defence for Air, Charles Power, inspects a graduating class of the EATS. In the foreground is Sergeant Willian Thomas Gunn, RNZAF.

AIR FORCE MUSEUM OF NEW ZEALAND, ALB910283033

LEFT | Bill Jordan talks with the crew of a Wellington bomber in No. 75 (NZ) Squadron.

AIR FORCE MUSEUM OF NEW ZEALAND, ALB021131B023

RIGHT | RNZAF student observers marching to class at the No. 4 Air Observers School in London, Ontario, in March 1941. The role of a bomber observer was to navigate the plane to the designated target and tell the pilot when to release the bombs.

AIR FORCE MUSEUM OF NEW ZEALAND, ALB910283003

BELOW | Time out for air force trainees at the Anzac Hospital Center in Chicago, Illinois. They are named, but the order is not clear: Charles Boss, Charles Cosson, Ross Hunter and Leonard Bint.

AIR FORCE MUSEUM OF NEW ZEALAND, ALB930201057

ABOVE | A singalong at the Piccadilly Hotel, New York, for New Zealand airmen in transit. Such events aided group bonding and helped to release pent-up tension.

AIR FORCE MUSEUM OF NEW ZEALAND, ALB88177048

LEFT | An early prototype of the Supermarine Spitfire, photographed in 1937. This very capable fighter plane, which went through several modifications during the war and was still in service in 1945, proved its worth for the RAF in the Battle of Britain in 1940.

AIR FORCE MUSEUM OF NEW ZEALAND, ALB8737528017

BELOW | A Spitfire from No. 74 Squadron over RAF Hornchurch in July 1939. Hornchurch, north-east of London, was home to a New Zealand convalescent hospital during the First World War.

AIR FORCE MUSEUM OF NEW ZEALAND, ALB8737528021

RIGHT | 'Mother and self.' Aircraftman George Dee poses with his mother in Waimate before sailing overseas. The smiles look somewhat strained and the image captures the anxiety most mothers felt about their sons going to war. Dee trained at the Bombing and Gunnery School in Alberta, Canada, and served as a bomb aimer with No. 156 Squadron Bomber Command in the United Kingdom.

AIR FORCE MUSEUM OF NEW ZEALAND, ALB040221001

ABOVE | 'Pilots of the No. 1 New Zealand Flight' pose in front of a Wellington bomber at RAF Station Marham in Norfolk, England, in August 1939. They are, from left: Flying Officer John Collins, Flight Lieutenant Charles Hunter, Flying Officer John 'Jack' Adams, Pilot Officer Trevor Freeman, Squadron Leader Cyril 'Cyrus' Kay, Wing Commander Maurice 'Buck' Buckley (commanding officer), Flying Officer Arthur Greenaway, Flight Lieutenant Aubrey Breckon, Flying Officer Neville Williams, Flying Officer Fred 'Popeye' Lucas, Flying Officer William Coleman and Pilot Officer Wilfred 'Bill' Williams. These 12 men remained in Britain to form the nucleus of No. 75 (NZ) Squadron.

AIR FORCE MUSEUM OF NEW ZEALAND, ALB8737528001

RIGHT | Members of No. 75 (NZ) Squadron examine a map before undertaking a long-range reconnaissance flight to Narvik, Norway, in April 1940. From left: Leading Aircraftman Edwin Williams, Flight Lieutenant Aubrey Breckon, Lieutenant Commander F. O. Howie (Royal Navy), Sergeant Robert Hughes (navigator), Pilot Officer Donald Harkness (second pilot) and Aircraftman Thomas Mumby (gunner observer). Breckon would be the pilot for this flight.

AIR FORCE MUSEUM OF NEW ZEALAND, PR9156

LEFT | A sergeant taking a sight from the cockpit of a Wellington bomber of No. 75 (NZ) Squadron before a flight to Narvik in April 1940.

AIR FORCE MUSEUM OF NEW ZEALAND, ALB0021131B056

BELOW | A Wellington bomber of No. 75 Squadron piloted by Aubrey Breckon taking off for Narvik in April 1940.

AIR FORCE MUSEUM OF NEW ZEALAND, ALB021131B058

ABOVE | 'Hurry up and wait', a process familiar to all soldiers. New Zealanders on the docks at Alexandria wait to board their troopships bound for Greece, although they did not then know their destination.

NATIONAL ARMY MUSEUM, 1992-982

| CHAPTER 3 |

Greece and Crete

According to one historian, 2 NZEF in the Middle East, which comprised a significant proportion of New Zealand's keenest and most capable young men, was 'the equivalent of twenty-five divisions of British troops; and in New Zealand's thinking the dangers which it encountered had to be considered on that scale'.[1] In early 1941, however, barely formed and assembled, and completely inexperienced — it had not even undertaken a single combined training exercise — the New Zealand Division of 2 NZEF was about to face a severe test.

In 1939, faced with aggressive German expansion, and at last recognising that Hitler could not be appeased, Britain and France issued territorial guarantees not only to Poland but also to Greece and Romania. On 28 October 1940 Italy invaded Greece and after suffering a series of humiliating defeats was forced to fall back into Albania. Hitler, concerned by this turn of events, needed to secure Europe's southern flank in preparation for his invasion of the Soviet Union. In response Germany began massing an army in Bulgaria that was clearly preparing to invade Greece and Yugoslavia. Aware of this build-up, the Greek government requested assistance from Britain. Winston Churchill, excited by prospects of a 'Balkan front' consisting of Greece, Yugoslavia and Turkey, directed that British theatre commander General Archibald Wavell cease military operations in the Western Desert and send a force to Greece. The die was cast for a military disaster.

Already overstretched in the region, Britain could send only a token force to Greece. Labelled W Force, after the commander General Sir Henry Maitland Wilson, it consisted of 6 Australian Division, the New Zealand Division, one brigade of tanks and five artillery regiments, and 80 RAF aircraft. All the military commanders involved, including Freyberg and the Australian commander Thomas Blamey, knew that W Force had no prospect of success but believed their governments had been told about the considerable risks involved. Freyberg wrote in his diary that the plan for the deployment to Greece 'violates every principle of war'.[2] He assumed that the New Zealand government had been consulted and had approved the use of the division. Freyberg's failure to inform the government of the dangers, and of his apprehensions, would almost cost him his command.

The advance parties of the New Zealand Division reached Athens on 7 March

1941, and during that month and into early April spent considerable time and effort preparing defensive positions along the Aliakmon Line in northern Greece. But then, on 6 April, came the German attack on both Greece and Yugoslavia with an overwhelming force consisting of two complete armies, numbering some 27 divisions, and over 1000 aircraft. Yugoslavia was quickly overrun, allowing the Germans to outflank the Allied positions. One German army advancing from Bulgaria soon sent the Greeks reeling back from the Metaxas Line in the north while the other advanced from Yugoslavia behind the Aliakmon Line. On 10 April British and German units clashed for the first time and many British units fought well. Others, however, did not: one Australian and one New Zealand battalion abandoned their positions in wild panic. W Force could only delay the inevitable and the Aliakmon Line had to be abandoned. For W Force, the campaign in Greece deteriorated into that most difficult and tiring phase of war, a protracted withdrawal. 'We were all rather fed up,' Kippenberger wrote later, 'and would have liked to have seen some German infantry, though we were not particularly keen on the A.F.V.s [armoured fighting vehicles].'[3]

For W Force, the campaign could only end one way: in defeat. This was known from the start: British military authorities in Cairo began planning for an evacuation from Greece even as the first troops landed there. Initially the naval planners believed they would be lucky to save more than a third of W Force,[4] but the Royal Navy did better than this. Over five days, from 22 to 26 April, in an operation reminiscent of Dunkirk the previous year, the Royal Navy evacuated 50,172 of the 62,212 men of W Force, mostly Australians and New Zealanders.[5] The majority of the almost 14,000 British casualties became prisoners of war, including 1614 New Zealanders.[6] (In fact 1856 were captured but 242 were wounded and could not be evacuated.) This was the first time such a large number of New Zealanders had been captured in an overseas military operation; about 500 were taken prisoner in the entire First World War.[7] Those captured in 1941 had long years of captivity, isolation and deprivation ahead of them. The Greek campaign achieved nothing and was a painful lesson in what not to do. Worse was to come in May 1941.

THE ISLAND OF CRETE lies in the eastern Mediterranean 160 kilometres from Greece. In 1941 it had a good port at Suda Bay and three airfields at Maleme, Heraklion and Retimo. The British, who recognised Crete's strategic value, had been allowed to garrison it since May 1940. However, with little military equipment to spare and their attention elsewhere, the British did almost nothing to prepare the island's defences against a German invasion.

In the evacuation from Greece, New Zealand 6 Brigade was fortunate to be taken

LEFT | Soldiers in Egypt bound for Greece board SS *Hellas*, which had a very rough Mediterranean crossing.

NATIONAL ARMY MUSEUM, 1993-1331

ABOVE | A group of New Zealand officers of 23 Battalion look pleased to be arriving in Greece.

NATIONAL ARMY MUSEUM, 2001-1316-1

RIGHT | The Greeks had not been expecting a British expeditionary force. They were overjoyed to see them and, according to the official history, gave all arrivals 'a spontaneous and tumultuous welcome'.

NATIONAL ARMY MUSEUM, 1989-543

RIGHT | Gunners of 25 Battery, 4 Field Regiment, enjoy a brew in their gun pit at Koukos, near Katerini, in April 1941. They would soon be on the move and eventually would have to destroy their guns.

COLLINS COLLECTION

ABOVE | Patients and staff at a hospital in Athens.

NATIONAL ARMY MUSEUM, 1992-1084-7

RIGHT | NCOs from 23 Battalion on a train leaving Athens.

NATIONAL ARMY MUSEUM, 2001-1316-1

directly to Alexandria. In order to rescue more soldiers more quickly, the Royal Navy offloaded New Zealand 4 and 5 Brigades and Australian 19 Brigade on Crete. It was expected that they would soon be on their way to Egypt. But evidence of an impending German invasion of Crete changed the situation and the Australians and New Zealanders joined the British garrison and 11,000 Greeks to defend the island. By the end of April there were more than 42,000 British Commonwealth and Greek soldiers on Crete.[8]

According to one New Zealand officer, the troops were 'bewildered' and lacked the necessary equipment to resist a German invasion.[9] There were not enough trucks, digging equipment, artillery, heavy machine guns or radios. Some obsolete tanks were sent, but the RAF aircraft left Crete just as the New Zealanders arrived. On 30 April, just a day after his arrival, Freyberg was appointed as the commanding officer of Creforce, the Allied troops on the island. It was not a task he wanted and nor did he think Crete could be held with the resources on hand. Freyberg had one great advantage, however. Thanks to Alan Turing and others at Bletchley Park breaking the German Enigma code, the British could read the encrypted signals of German Ultra intelligence, which meant Freyberg knew almost every detail of the enemy's invasion plans for Crete, including when the attack would be made. Some have questioned whether Freyberg had the intellect or skill to make proper use of this information. They claim that he worried too much about a seaborne invasion and did not commit sufficient resources to defending the airfields.[10]

There is a strong element of hindsight in these criticisms. Dan Davin, author of the official history of the Crete campaign, concluded after intensive investigation that 'the preliminary measures taken on the island in May were reasonable in their character and as effective as could fairly be expected'.[11] Freyberg's orders were also clear. The airfields and port were to be defended to the utmost, units were to offer mutual support if one became hard-pressed, and if positions were lost an immediate counterattack was necessary.

On 25 April 1941 Hitler issued Directive 28 by which the Luftwaffe was to occupy Crete in Operation 'Merkur' (Mercury). Three parachute regiments of 7 Air Division would spearhead the assault. The plan was to drop paratroop and glider-borne infantry in two waves — at Maleme and Suda Bay in the morning, followed by drops at Retimo and Heraklion in the afternoon. By nightfall all three airfields should be in their hands and reinforcements would arrive by sea and air the next day. German intelligence was poor; they expected little opposition. The German invasion of Crete began at dawn on 20 May 1941. Watching the parachutes descend, Freyberg remarked to a young British officer breakfasting with him, 'They are dead on time.'[12] Everywhere parachutists landed that day they suffered appalling losses and for such little gain that, by the evening, General Kurt Student, the German commander of 11 Air Corps, was convinced that the battle was lost. The German paratroopers did, though, capture ground and consolidate their positions in one significant location: the Tavronitis riverbed west of Maleme airfield. They then pressed on towards Maleme. When a counterattack failed to halt the paratroopers, the New Zealand battalion commander panicked. When his request for assistance was not answered, he withdrew his troops from Maleme airfield and the vital hill overlooking it. It was the turning point of the battle.

By dawn on 21 May, the Germans had secured Maleme and that afternoon they landed 40 transport planes, which brought in vital reinforcements and supplies. A New Zealand counterattack at dawn on 22 May was poorly planned and executed. It was a case of too little, too late. In the words of Ian McGibbon, 'the failure of the counterattack decided the fate of Creforce'.[13] The battle lasted for another eight days, with some ferocious counterattacks, but the situation could not be redeemed. The fight for Crete now deteriorated into one long, arduous withdrawal over the White Mountains to the evacuation beaches at the small fishing village of Sfakia on the south coast. On 27 May Freyberg requested and received permission to evacuate the island. As he told Wavell in Cairo, 'The troops had reached the limit of their endurance and the position was hopeless.'[14] Yet again the Royal Navy would need to come to the rescue.

FROM 28 MAY to 1 June 1941 some 17,000 men were evacuated from Crete, but over 12,000 Allied soldiers, including 2180 New Zealanders, 488 of whom were wounded, had to be left behind to be taken prisoner.[15] Crete was nearly Germany's first land defeat of the war. This was avoided primarily because of a series of New Zealand command failures. The Battle for Crete also affected the Germans. According to Student, Crete was 'the grave of the parachutist'. Hitler was determined never again to risk these elite troops in a major air drop.[16]

The New Zealand Division was not ready and should never have been used in the Greek campaign. As Freyberg later acknowledged, 'Greece was a disaster, Crete was a tragedy.'[17] Both campaigns were part of a long chain of disasters for British forces that persisted well into 1942. Because of poor planning, faulty leadership, lack of decent equipment and inadequate training, the New Zealanders had experienced the bitter taste of defeat. In the months ahead it was a taste that became all too familiar.

RIGHT | Soldiers from 23 Battalion take a break at a pass in Greece. Both beer and tea are available.

NATIONAL ARMY MUSEUM, 2001-1316-1

BELOW | Early in the Greek campaign, New Zealand soldiers enjoyed being in terrain that reminded them of home.

NATIONAL ARMY MUSEUM, 1992-1174

ABOVE | Soldiers from 18 Battalion in wintry conditions.

NATIONAL ARMY MUSEUM, 1992-1084-7

RIGHT | New Zealand soldiers endure a strenuous journey in the snow.

NATIONAL ARMY MUSEUM, 1993-1331

LEFT | By the time the New Zealanders arrived in Greece, the country had been at war against Italy since the end of October 1940. Here a group of captured Italian soldiers are being marched to their detention camp, but they would not be prisoners of war for much longer.

NATIONAL ARMY MUSEUM, 2004-267-2

RIGHT | A New Zealand ambulance passes Greek civilians, most of them women, working on the road.

NATIONAL ARMY MUSEUM, 2005-587-2

BELOW | A New Zealand Field Ambulance reception centre. The trees provide some cover from aerial observation.

NATIONAL ARMY MUSEUM, 2005-587-2

RIGHT | A regimental aid post (RAP) in Greece. The New Zealand medics offered treatment to locals if time and resources permitted.

NATIONAL ARMY MUSEUM, 2001-1117-1

BELOW | Lunchtime in the field for this section of New Zealand gunners.

NATIONAL ARMY MUSEUM, 2001-720

ABOVE | The generals in Greece: from left, Australian commander General Thomas Blamey, W Force commander Field Marshal Henry 'Jumbo' Wilson, and New Zealand's General Bernard Freyberg.

NATIONAL ARMY MUSEUM, 2004-267-2

LEFT | General Freyberg snatches a quick shave during the campaign.

NATIONAL ARMY MUSEUM, 2004-267-2

ABOVE | The Servia Pass is bombed by the Germans in early April 1941.

NATIONAL ARMY MUSEUM, 2001-720

RIGHT | The ruins of the junction town of Larissa, through which most New Zealand soldiers passed on their way to the coastal ports.

NATIONAL ARMY MUSEUM, 1993-1331

RIGHT | At Trikala, Greece, on 15 April 1941, gunners of A Troop, 25 Battery, 4 Field Regiment stand among the wreckage of a downed Ju-87 Stuka dive-bomber, which had crashed close to their gun position. Surprisingly the 19-year-old pilot, Walter Seeliger, and his gunner, Kurt Friedrich, survived and were captured.

COLLINS COLLECTION

ABOVE | During the long withdrawal from Greece, exhausted soldiers take the opportunity to rest in an olive grove.

NATIONAL ARMY MUSEUM, 2006-267-2

LEFT | The soldier in this photograph, taken on his last day in Greece, is Colwyn Smith of the Divisional Intelligence Section.

NATIONAL ARMY MUSEUM, 2004-672-2

LEFT | Sunken British ships in a Greek harbour. The vessel to the right, SS *Clan Cumming*, was hit by a German mine and sank in April 1941.

NATIONAL ARMY MUSEUM, 1992-1174

ABOVE | On 24 April German paratroopers captured the Corinth Canal in Greece in a daring airborne operation, though they were too late to prevent the destruction of the bridge, which had been wired for demolition by the British. The Germans were cutting the fuses to the explosives when the charges exploded.

NATIONAL ARMY MUSEUM, 1992-184-7

LEFT | German Mountain troops at Naples on their way to Greece.

NATIONAL ARMY MUSEUM, 1991-770

ABOVE | Two New Zealand soldiers evacuated from Greece on board the light cruiser *Ajax*, which had fought with *Achilles* at the River Plate in 1939. In April 1941, over several trips, it evacuated soldiers from 4 Brigade and 5 Brigade to Crete.

NATIONAL ARMY MUSEUM, 1992-1084-7

LEFT | The lucky ones — a crowded ship returns soldiers to Egypt after the Greek campaign.

NATIONAL ARMY MUSEUM, 2005-587-2

ABOVE | Digging in on Crete. The locals assisted whenever they could and later paid a high price for their involvement.

NATIONAL ARMY MUSEUM, 1992-1174

RIGHT | New Zealand soldiers, bewildered and short of vital equipment, prepare for stand-to on Crete.

NATIONAL ARMY MUSEUM, 2004-267-2

RIGHT | A German air raid at Suda Bay. For almost three weeks Crete was constantly attacked from the air and the RAF could offer no resistance.

NATIONAL ARMY MUSEUM, 2004-267-2

BELOW | New Zealand soldiers prepare for operations in Crete.

NATIONAL ARMY MUSEUM, 1992-1174

ABOVE | Early morning on 20 May 1941 and the airborne invasion of Crete is under way. This was the drop over Suda Bay. A German transport plane has been hit.

NATIONAL ARMY MUSEUM, 1991-770

RIGHT | German Mountain troops boarding a transport plane on their way to Crete.

NATIONAL ARMY MUSEUM, 1992-1084-7

ABOVE | German parachutes left in the trees of Crete. If they were lucky enough to land unscathed, paratroopers had to fight for their lives.

NATIONAL ARMY MUSEUM, 1993-1352-1

RIGHT | And still they come — more paratroopers are dropped over Crete. Dan Davin, the New Zealand official historian of the Crete campaign, wrote that the sight of the paratroopers was 'inexpressibly sinister. For each man dangling carried a death, his own if not another's.'

NATIONAL ARMY MUSEUM, 2006-49

ABOVE | A German Ju-52 transport plane lands in Crete. The capture and holding of the Maleme airfield by the Germans was the turning point of the battle.

NATIONAL ARMY MUSEUM, 1993-1352-1

LEFT | British soldiers taken prisoner in Crete. Over 2000 of them were New Zealanders.

NATIONAL ARMY MUSEUM, 1992-1084-8

ABOVE | Tom Goodall of 22 Battalion made it to Egypt after the Crete battle but became a prisoner of war in 1942.

NATIONAL ARMY MUSEUM, 2000-723-1

ABOVE | Armourers prepare to load bombs onto a Wellington bomber of No. 75 (NZ) Squadron. This photograph was most likely taken at RAF Station Feltwell in Norfolk around December 1940.

AIR FORCE MUSEUM OF NEW ZEALAND, MUS090236

| CHAPTER 4 |

The Air War over Europe

In the First World War aircraft had proved useful, especially in their reconnaissance role, but they were never decisive in the outcome of a battle. In the Second World War, however, aircraft and controlling the airspace were essential to the outcome of any clash of arms. As Professor Sir Richard Overy notes, 'Air power emerged as a major strategic element . . . and for the rest of the twentieth century remained so.'[1] Victory was impossible without air superiority over the enemy. There was one exception to this. From 1943 to 1945 in Italy the Allied air forces ruled the skies, but the Germans still fought doggedly and inflicted several serious defeats.

When New Zealand joined the war in September 1939 the RNZAF was just two years old. It was also small, consisting of just 756 full-time personnel and 404 Territorial Force members.[2] Most of its 102 aircraft were second hand and near obsolete. Only 14 were relatively new: five Oxford twin-engine trainers and nine Vickers Vildebeest single-engine biplane bombers.[3] This situation was rapidly transformed as the RNZAF launched a large-scale recruitment campaign from the moment war was declared. It established new flying training schools at Taieri, Harewood, New Plymouth, Levin and Whenuapai, and an Air Gunners and Observers School at Ohakea. The Territorial Force squadrons were mobilised. From fewer than 1200 men and 102 planes in 1939, the RNZAF reached a peak strength of 42,000 personnel in June 1944, with 1336 operational aircraft. There were 24 RNZAF squadrons in the Pacific theatre. New Zealand has never had as many planes and personnel in its air force since that time.[4]

A vital part of New Zealand's contribution to the air war was the 12,000 men it sent to serve with the RAF in Europe. New Zealanders served in all the campaigns fought by the RAF and in all three of its operational branches: Bomber, Fighter and Coastal Commands.

In September 1939 some 550 New Zealanders were serving in the RAF. During the war years more than 7000 New Zealand airmen joined the RAF via training in Canada; many, trained in New Zealand, went directly to Britain. Around 10 per cent of this latter group served in one of seven designated 'New Zealand' squadrons in the RAF: 75 and 487 in Bomber Command; 485 (Spitfires), 486 (Hurricanes) and 488 (Buffaloes, Hurricanes, Beaufighters, Mosquitoes) in Fighter Command;

and 489 and 490 Squadrons in Coastal Command. According to historian Christina Goulter, 12,078 New Zealanders served in the RAF during the war, of whom 3285 (27 per cent) died. Another 500 New Zealanders with the RAF became prisoners of war.[5]

OF THE 125,000 personnel in Bomber Command, 47,000 were killed in action, 8000 died during training and 2000 ground crew died during the war. This was a fatality rate of 46 per cent. A further 10,000 became prisoners of war, 4000 returned wounded from raids and as many again were injured in accidents. This made total aircrew casualties a terrible 75,000, almost 60 per cent of all those in Bomber Command. Of the 6000 New Zealanders who served with Bomber Command, almost one in three were killed in action. This was the highest ratio of any New Zealand service during the war. Of the some 1850 who lie in some corner of a foreign field that is forever New Zealand, 400 have no known grave.[6] The first of the dominion squadrons, No. 75, lost so many men on its numerous operations over five years that it became known as 'the chop squadron'. Its members suffered nearly a quarter of all New Zealand deaths in Bomber Command.[7]

The actions of Bomber Command have been controversial ever since the war ended. Two aspects have been hotly contested: its strategic value and the morality of destroying most of Germany's towns and cities, with severe civilian casualties. What has never been denied, though, is the 'skill, courage and determination' of the aircrews involved and the great risks they took doing their duty.[8]

During the Battle of Britain in 1940, the first air clash to have truly decisive results, New Zealanders formed the third-largest national group in Fighter Command, after the Poles and the English.[9] Two New Zealand squadrons also served with Coastal Command in anti-shipping, anti-submarine and reconnaissance roles, which have 'received little academic and popular attention'. The pilots themselves certainly believed that their story had been 'overshadowed by the perhaps more glamorous exploits of Bomber and Fighter Commands'.[10]

Winning the air war over Britain and then over Europe was essential to defeating Nazi Germany. Though relatively small in number, the New Zealanders who flew with the RAF 'could rightly claim that they played a significant part in the air war — a war which proved decisive in the overall Allied effort'.[11]

ABOVE | 'The Boys' — No. 3 Squadron pilots at RAF Station Castletown in Scotland, November 1940. The several New Zealanders in this squadron included Herbert Mitchell, whose album provided this and other images. Mitchell, from Reefton, was listed as missing on operations on 12 May 1942.

AIR FORCE MUSEUM OF NEW ZEALAND, ALB863701033

LEFT | 'Gen talk': pilots of No. 3 Squadron share information at RAF Castletown in November 1940.

AIR FORCE MUSEUM OF NEW ZEALAND, ALB863701035

RIGHT | Pilot Officer J. Allen (right) and Observer Flight Sergeant W. M. Patterson examine the remains of a German Ju-88 bomber they shot down on the night of 18–19 April 1940.

AIR FORCE MUSEUM OF NEW ZEALAND, ALB112149006

BELOW | In the 1940 Battle of Britain, Polish pilots made up the second largest national group, after Britain and just ahead of New Zealand. Sergeant Josef Biel (left) and Sergeant Pawel Gallus of No. 3 Squadron in front of a Hurricane fighter at RAF Station Castletown in November 1940.

AIR FORCE MUSEUM OF NEW ZEALAND, ALB863701034

RIGHT | '1941, Les ready for high altitude', reads the annotation on this photograph of Flight Sergeant Leslie Russell of No. 264 Squadron beside his Defiant fighter. The Defiant failed as a day fighter because of its lack of forward firing capability, and suffered huge casualties because of this. Like the Me 110, it was more successful as a night fighter.

AIR FORCE MUSEUM OF NEW ZEALAND, ALB140144322

ABOVE | Flight Sergeant Colin Hamilton from Wellington standing in front of a Hurricane fighter of No. 260 Squadron in 1942. Although overshadowed somewhat by the more glamorous Spitfire, the Hurricane was the workhorse of Fighter Command until early 1942. Hamilton was killed on 15 August 1943 while fighting in the Mediterranean theatre.

AIR FORCE MUSEUM OF NEW ZEALAND, ALB941291092

LEFT | An unknown pilot of No. 260 Squadron gives the V for victory sign from the cockpit of his Hurricane.

AIR FORCE MUSEUM OF NEW ZEALAND, ALB863701094

RIGHT | A flight of No. 260 Squadron in front of a Hurricane at RAF Base Skitten in Scotland in 1941.

AIR FORCE MUSEUM OF NEW ZEALAND, ALB863701068

ABOVE | Sergeant Allan Box (left) in discussion with Sergeant James Ward of Whanganui at the back of their Wellington bomber in August 1941. In July, while Ward climbed out on the wing of his damaged bomber to extinguish a fire, Box shot down the ME110 night fighter that had caused the damage. Ward was awarded the Victoria Cross, Box a Distinguished Flying Medal. On 15 September 1941, Ward was killed in action after the plane he was piloting was shot down by a German night fighter. Box survived the war and died in 1987.

AIR FORCE MUSEUM OF NEW ZEALAND, ALB021131B005

ABOVE | Taking the bus to work. Cheerful No. 75 (NZ) Squadron personnel about to be taken to their waiting aircraft.

AIR FORCE MUSEUM OF NEW ZEALAND, ALB933373006

LEFT | Ground crew loading up a Wellington bomber.

AIR FORCE MUSEUM OF NEW ZEALAND, ALB098954017

RIGHT | Leonard Hewitt at the controls of a Wellington bomber landing in Egypt in 1941. Hewitt had been with No. 75 (NZ) Squadron in Britain but was now part of No. 37 Squadron.

AIR FORCE MUSEUM OF NEW ZEALAND, ALB0224349168

LEFT | In this early 1942 photograph of his operational crew in the Western Desert, Leonard Hewitt is standing on the right.

AIR FORCE MUSEUM OF NEW ZEALAND, ALB0224349278

LEFT | Leonard Hewitt, centre, with two other members of No. 37 Squadron, in a sandbagged bomb shelter in Egypt. They are named as MacFarlane and Nesbit.

AIR FORCE MUSEUM OF NEW ZEALAND, ALB0224349279

BELOW | Oil tanks exploding after an air raid on Pembroke Dock, Wales, in 1941.

AIR FORCE MUSEUM OF NEW ZEALAND, ALB13205100

BELOW | A London flying boat of No. 204 Squadron explodes in May 1941, probably at RAF Station Sullom Voe in Scotland.

AIR FORCE MUSEUM OF NEW ZEALAND, ALB13205089

RIGHT | In August 1941 a No. 204 Squadron Catalina practises a low bombing run at Invergordon in Scotland.

AIR FORCE MUSEUM OF NEW ZEALAND, ALB13205092, ALB13205093

LEFT | Loading a 450-pound depth charge at Sullom Voe in April 1941. From left: Garry Craig, Les Cordes, Lofty Knott and Sergeant Fearns of No. 204 Squadron.

AIR FORCE MUSEUM OF NEW ZEALAND, ALB13205087

LEFT | Of the six crew boarding a No. 75 (NZ) Squadron Wellington bomber at RAF Station Feltwell, three were New Zealanders: Franklyn Cran, Maurice Bell and Claude Harris. This aircraft, and its entire crew, were lost during a raid on Lubeck on the night of 28–29 March 1942.

AIR FORCE MUSEUM OF NEW ZEALAND, ALB001252026

BELOW | The aftermath of the bombing of London in 1941. This area was behind High Holborn.

AIR FORCE MUSEUM OF NEW ZEALAND, ALB140144153

RIGHT | A group of No. 486 (NZ) Squadron personnel with one of the Hurricane fighters at RAF Station Wittering in 1942.

AIR FORCE MUSEUM OF NEW ZEALAND, 1988-1176-13

BELOW | Stanley Browne in his No. 485 (NZ) Squadron Spitfire, probably at RAF Station Kenley in 1942. Browne, a medical student from Wellington, survived being shot down over France to command the squadron at the end of the war. He achieved ace fighter status by downing more than five enemy aircraft.

AIR FORCE MUSEUM OF NEW ZEALAND, ALB090071008

LEFT | Squadron Leader Reginald Grant, DFC and Bar, DFM, from Woodville, in the cockpit of his Spitfire fighter. Grant, who rose through the ranks to command No. 485 (NZ) Squadron, was killed on operations on 28 February 1944, while heading 122 Wing.

AIR FORCE MUSEUM OF NEW ZEALAND, ALB090071004

ABOVE | Flight Lieutenant William (Bill) Crawford-Compton of Invercargill standing on the wing of a No. 485 (NZ) Squadron Spitfire at RAF Station Kenley in 1942. After the war, Crawford-Compton remained in the RAF and reached the rank of air vice-marshal.

AIR FORCE MUSEUM OF NEW ZEALAND, ALB090071033

RIGHT | A No. 485 (NZ) Squadron Spitfire flown by William Crawford-Compton shows its battle scars. Partly obscured is a caricature of Hitler in a frying pan over a burning swastika.

AIR FORCE MUSEUM OF NEW ZEALAND, ALB090071012

ABOVE | Armourers of No. 486 (NZ) Squadron service the machine guns of the fighter planes. The pictures of women on the wall, probably torn out of magazines, are by the doyen of pin-up art, Peruvian American Alberto Vargas.

AIR FORCE MUSEUM OF NEW ZEALAND, 2018-026-12

RIGHT | There was no mistaking that No. 486 was a New Zealand squadron.

AIR FORCE MUSEUM OF NEW ZEALAND, 2018-026.4

ABOVE | At RAF Station Wittering in 1942, No. 486 (NZ) Squadron armourers prepare a Hurricane fighter for operations. They are named, from left, R. D. Fairbrother, R. K. Bartlett, J. P. Walton and K. K. Moore, with P. Brown on the wing. Bartlett was killed in 1944.

AIR FORCE MUSEUM OF NEW ZEALAND, 1988-1176.14B

RIGHT | Pilots of No. 485 (NZ) Squadron, with a Spitfire in the background, at RAF Station Kenley.

AIR FORCE MUSEUM OF NEW ZEALAND, ALB090071026

BELOW | The pilots of No. 485 (NZ) Squadron photographed in front of a Spitfire in 1942 are, standing from left, J. D. (Jack) Rae, A. G. Shaw, D. Russell, P. H. Gaskin, Bill Crawford-Compton and Evan Mackie. In front are D. G. E. Brown and D. T. Clouston.

AIR FORCE MUSEUM OF NEW ZEALAND, ALB090071031

ABOVE | Pilots from No. 485 (NZ) Squadron in their dispersal hut at RAF Station Kenley.

AIR FORCE MUSEUM OF NEW ZEALAND, ALB090071024

RIGHT | Pilots of No. 485 (NZ) Squadron with Alan Deere's dog, Steve. They are named, from left, E. D. Mackie, Sgt. Shaw, D. T. Clouston, J. G. Pattison and J. J. Palmer. The photograph was taken in 1942.

AIR FORCE MUSEUM OF NEW ZEALAND, ALB090071025_1

LEFT | No. 485 (NZ) Squadron pilots in February 1942, just after completing a successful action over German battleships in the English Channel. From left: H. N. Sweetman, D. T. Clouston, B. E. Gibbs (obscured), R. J. C. Grant, M. M. Shane (obscured), E. P. Wells, M. McNeill (obscured), W. V. Crawford-Compton, J. M. Checketts and R. W. Baker. Flying ace Johnny Checketts would be awarded both the DSO and a DFC.

AIR FORCE MUSEUM OF NEW ZEALAND, ALB090071001

LEFT | Bomber aircrew receiving a briefing from an intelligence officer before an operation. Despite extensive preparations, aerial bombing was not always accurate.

AIR FORCE MUSEUM OF NEW ZEALAND, ALB098954016

BELOW | 'Take-off Time'. Members of No. 75 (NZ) Squadron ready to board their Wellington bomber at RAF Station Feltwell in June 1942. The pilot was Allen Fraser (second from right) from Palmerston North, who was killed in action, aged only 24, on an operation to Emden on 21 June 1942.

AIR FORCE MUSEUM OF NEW ZEALAND, ALB030841047

RIGHT | Crews from No. 75 (NZ) Squadron rest on the grass, their Lancaster bombers lined up behind them.

AIR FORCE MUSEUM OF NEW ZEALAND, ALB140143254

ABOVE | This Short Stirling bomber, ready for take-off in 1942, is most likely from No. 218 Squadron. The Stirling, though manoeuvrable, had a low operational ceiling and limited bomb load. It was replaced in late 1943. New Zealand war hero Phil Lamason served as a pilot in No. 218 Squadron. Lamason was the senior officer of a group of 168 Allied airmen sent to Buchenwald concentration camp in 1944.

AIR FORCE MUSEUM OF NEW ZEALAND, ALB140144208

BELOW | According to the annotation on this 1941 photo of airmen in full flying kit under the nose of a Hudson bomber, '2 of the 6 were killed by 1945'.

AIR FORCE MUSEUM OF NEW ZEALAND, ALB875584060

LEFT | Air Gunner Ronald Allen of No. 75 (NZ) Squadron was only 20 years old when he was killed on operations over Germany on 29 March 1942.

AIR FORCE MUSEUM OF NEW ZEALAND, ALB001252005

ABOVE | Pall-bearers carrying the coffins of three members of No. 218 Squadron who died when their Stirling bomber crashed near Marham on 25 September 1942, after returning, severely damaged, from an operation over Wilhelmshaven. Two were New Zealanders, Flight Sergeants Sidney Ives and Ernest (Trevor) Pellow. The Australian pilot was Flying Officer J. C. Frankcombe, not Franklin as it says on the photograph. Five of the seven-man crew lost their lives.

AIR FORCE MUSEUM OF NEW ZEALAND, ALB140144236

LEFT | Members of No. 218 Squadron farewell their three colleagues.

AIR FORCE MUSEUM OF NEW ZEALAND, ALB140144235

RIGHT | Flying Officer Robert Herron, from Dunedin, in a No. 75 (NZ) Squadron Lancaster. Despite the bomber's large frame, conditions on board were still cramped. Herron was listed as missing on operations on 28 April 1944.

AIR FORCE MUSEUM OF NEW ZEALAND, ALB021131C020

ABOVE | An unusual image of a No. 75 (NZ) Squadron Lancaster viewed from the full bomb bay of another Lancaster.

AIR FORCE MUSEUM OF NEW ZEALAND, ALB021131C052

RIGHT | Flight Lieutenant Richard (Dick) Broadbent (left) and Wing Commander E. P. Wells inspect a badly damaged No. 75 (NZ) Squadron Stirling bomber.

AIR FORCE MUSEUM OF NEW ZEALAND, ALB021131A008

RIGHT | A pilot's eye view of No. 75 (NZ) Squadron Lancasters on operations.

AIR FORCE MUSEUM OF NEW ZEALAND, ALB021131C067

ABOVE | The most vulnerable member of the aircrew. Tail gunner Flight Sergeant George Gardner from Whanganui demonstrates his fighting position in a No. 622 Squadron Lancaster bomber. Gardner was listed as missing on operations on 8 June 1944.

AIR FORCE MUSEUM OF NEW ZEALAND, ALB021131C007

ABOVE | Two air crew test the microphones in their helmets before flying.

AIR FORCE MUSEUM OF NEW ZEALAND, ALB88125123B083

BELOW | A wireless operator of No. 75 (NZ) Squadron at work.

AIR FORCE MUSEUM OF NEW ZEALAND, ALB021131C005

ABOVE | The strain of endless missions is clear on the faces of a No. 75 (NZ) Squadron bomber crew preparing to depart for a bombing raid over Germany.

AIR FORCE MUSEUM OF NEW ZEALAND, ALB021131B013

RIGHT | The crew of No. 75 (NZ) Squadron Wellington AA-D relaxing on the grass at Feltwell in June 1942.

AIR FORCE MUSEUM OF NEW ZEALAND, MUS0112911

ABOVE | Sergeant Edward Sadler, who disappeared in 'British Strength' over Burma on 26 March 1943, in the cockpit of a No. 67 Squadron Hurricane. This is a later version of the aircraft, armed with four cannons, rather than eight machine guns.

AIR FORCE MUSEUM OF NEW ZEALAND, ALB971526270

RIGHT | This pilot in the cockpit of a No. 125 Squadron Beaufighter, photographed in 1942, has a picture of Popeye on his inflatable life jacket, commonly known as a Mae West.

AIR FORCE MUSEUM OF NEW ZEALAND, ALB10051117065

LEFT | A member of the ground crew of No. 3 Squadron prepares a Hurricane fighter for a flight.

AIR FORCE MUSEUM OF NEW ZEALAND, ALB863701036

RIGHT | 'All my own work', pilot D. H. Mann has written on the back of this photograph of a No. 5 Operational Training Unit Hampden with a collapsed undercarriage. The accident happened at RAF Station Maghaberry, Northern Ireland, in January 1943.

AIR FORCE MUSEUM OF NEW ZEALAND, ALB901809029

LEFT | Written on this 'cookie' bomb being loaded onto a Lancaster bomber is: 'To Hitler. 4000lb greetings from the Southgate Gang.'

AIR FORCE MUSEUM OF NEW ZEALAND, ALB962381002

RIGHT | The Avro Lancaster could carry a massive payload, in this case a 12,000-pound blockbuster bomb. It became the workhorse of the RAF nightly bombing raids over Germany.

AIR FORCE MUSEUM OF NEW ZEALAND, ALB112149016

ABOVE | No. 226 Squadron bomb the Saint-Ghislain marshalling yards in Belgium on 26 April 1944. The photograph was taken from the bombing height of 10,000 feet.

AIR FORCE MUSEUM OF NEW ZEALAND, ALB80172CK062

ABOVE | The B-25 Mitchell 'T for Tommy' and crew after completing their 100th bombing operation. The aircraft was part of No. 226 Squadron of the 2nd Tactical Air Force. New Zealander Flight Lieutenant Stanley Wilks was a member of the crew and the pilot was Australian.

AIR FORCE MUSEUM OF NEW ZEALAND, ALB80172CK038

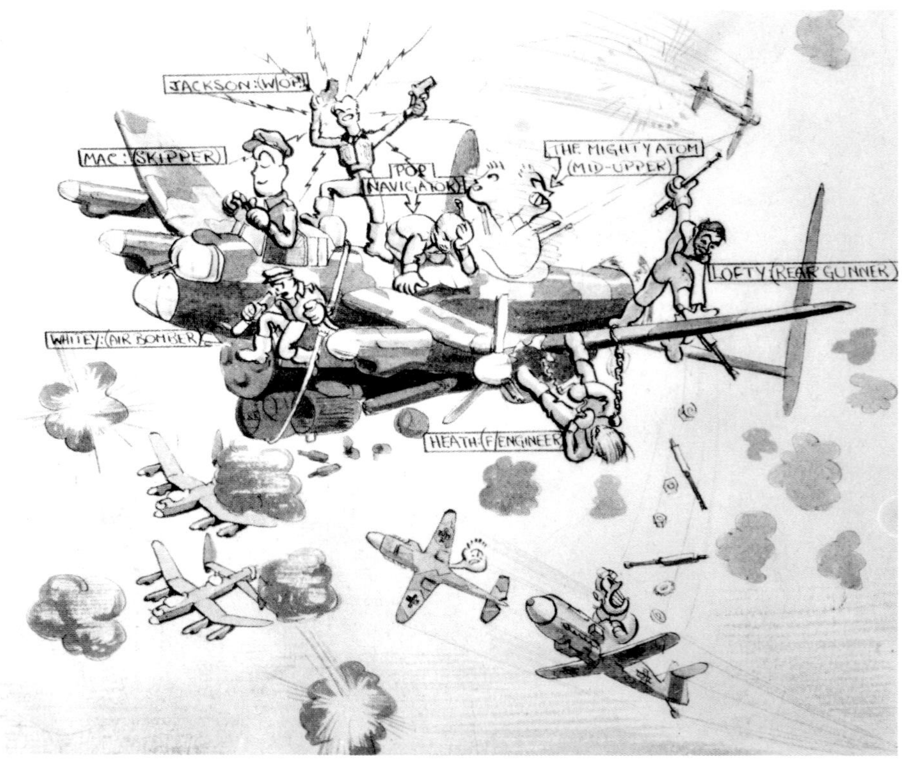

RIGHT | Cartoon of No. 115 Squadron Lancaster in flight, with all the crew doing their jobs. The Air Bomber 'Whitey' was Flying Officer Ivan (Jack) Williamson from Mossburn, who was awarded the Distinguished Flying Cross in July 1945.

AIR FORCE MUSEUM OF NEW ZEALAND, ALB962381001

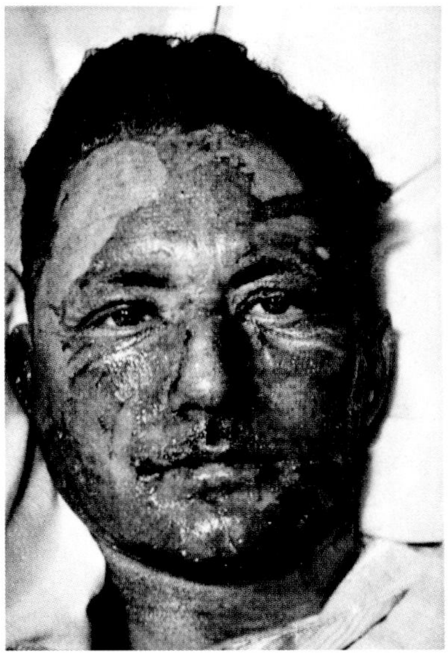

ABOVE | Flying Officer Ivan Williamson recovering in Ely Hospital after the crash of a No. 115 Squadron Lancaster bomber. Six crew members died, three of them New Zealanders; Williamson was one of two survivors. His facial burns were treated by pioneering New Zealand surgeon Sir Archibald McIndoe. As Williamson wrote, the burns ward at Ely 'was a "shocking" place in these days; strong stomachs very necessary'.

AIR FORCE MUSEUM OF NEW ZEALAND, ALB962381018

ABOVE | Mitchell bombers of No. 98 Squadron dropping bombs on an unknown target in 1944. Just below the cockpit and behind the aircraft code letter is the City of Derby coat of arms.

AIR FORCE MUSEUM OF NEW ZEALAND, ALB010711B284

BELOW | Smoke rises over Emmerich, Germany, after a bombing raid by No. 75 (NZ) Squadron. This photograph was taken on 7 October 1944.

AIR FORCE MUSEUM OF NEW ZEALAND, ALB021131D002

RIGHT | An aerial view of the merchant ship *Vim-Norge* taken during a No. 489 (NZ) Squadron anti-shipping patrol in August 1944.

AIR FORCE MUSEUM OF NEW ZEALAND, ALB901809085

LEFT | An aerial view of a downed No. 489 (NZ) Squadron crew in October 1944. The two men, D. H. Mann and Don Kennedy, spent nine days adrift in these life rafts before being rescued.

AIR FORCE MUSEUM OF NEW ZEALAND, 1990-180-31BI

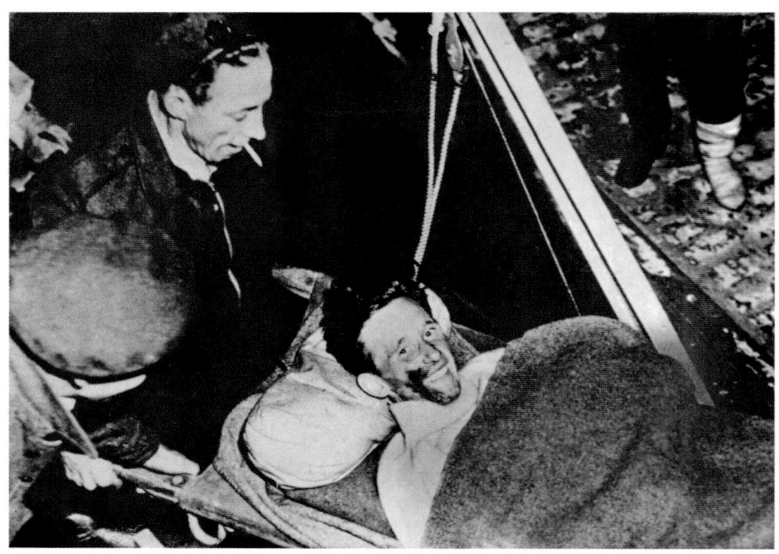

RIGHT | Navigator Don Kennedy looks very relieved to be rescued.

AIR FORCE MUSEUM OF NEW ZEALAND, ALB901809086

LEFT | A ground crew cleans the cannon of a Spitfire fighter in 1944, when these aircraft were being used to intercept V1 flying bombs, the first of Hitler's 'Wonder Weapons'.

AIR FORCE MUSEUM OF NEW ZEALAND, ALB112149010

RIGHT | Lancaster 'N for Nan' chalks up its 100th operational trip for Bomber Command. It was only the second heavy bomber, after 'S for Sugar', to complete a century of operations by mid-1944.

AIR FORCE MUSEUM OF NEW ZEALAND, ALB112149002

LEFT | This ground-to-air view of a large formation of Dakotas with escorts over the Dutch city of Arnhem was taken just moments before the paratroopers jumped. The paratroopers seized the bridge over the Rhine and held it for several days, but they were eventually overwhelmed. The fighter planes escorting the Dakota transports are part of the 2nd Tactical Air Force (2 TAF) commanded by the New Zealander Air Vice-Marshal Arthur Coningham.

AIR FORCE MUSEUM OF NEW ZEALAND, ALB031891182

BELOW | Lancaster 'C for Charlie' of No. 75 (NZ) Squadron Lancaster heading for the German town of Osterfeld on 22 February 1945.

AIR FORCE MUSEUM OF NEW ZEALAND, ALB33373002

ABOVE | Members of No. 75 (NZ) Squadron planning an operation in 1945 are, from left, Flying Officer Baker, Squadron Leader Charles Ormerod and Flying Officer Parsons.

AIR FORCE MUSEUM OF NEW ZEALAND, ALB021131C016

ABOVE | Returning from a night raid over Germany in February 1945, this Lancaster bomber overshot the runway and crashed into the village of Sutton. Surprisingly, no one was injured.

AIR FORCE MUSEUM OF NEW ZEALAND, ALB140145026

RIGHT | Bombs from No. 75 (NZ) Squadron fall on Ludwigshafen, Germany, on 5 January 1945.

AIR FORCE MUSEUM OF NEW ZEALAND, ALB0224347012

RIGHT | The largest bombs used by the RAF in the war, 'Grand Slams' were 35 feet 5 inches long and weighed 22,000 pounds. Designed to penetrate underground structures and precision targets requiring deep penetration, they caused massive damage. One bomb could make a crater 120 feet wide and more than 35 feet deep. They were first employed in March 1945.

AIR FORCE MUSEUM OF NEW ZEALAND, ALB112149009

BELOW | This Lancaster, whose nose art states 'Getsum Inn', came to a sad end after completing 92 successful missions.

AIR FORCE MUSEUM OF NEW ZEALAND, ALB140145028

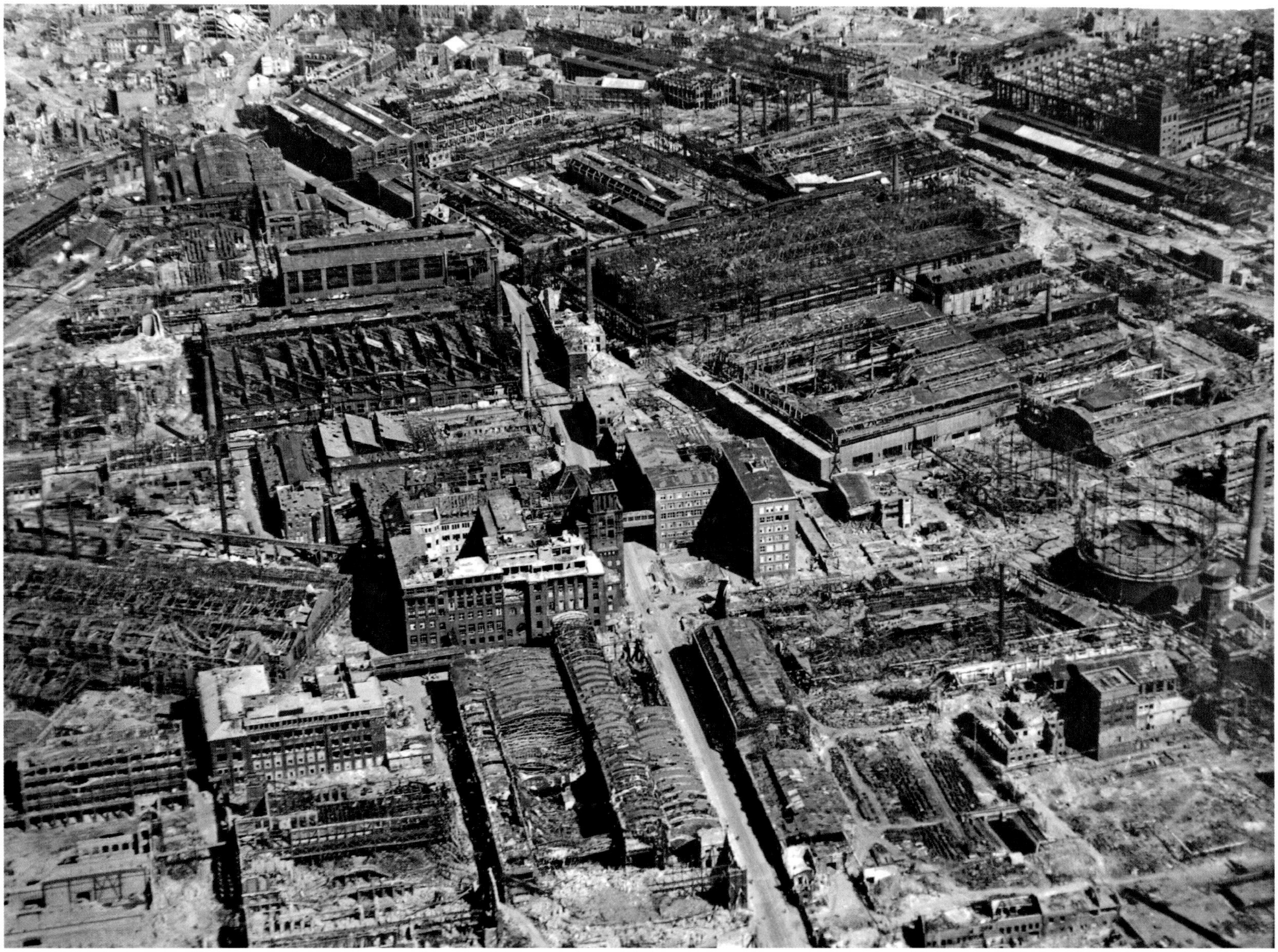

ABOVE | An aerial view of the damaged Krupps factory at Essen, Germany, taken in 1945. Essen was a primary target for Bomber Command. It was bombed for the final time in March 1945, when over 200 heavy bombers struck the coking plant and marshalling yards.

AIR FORCE MUSEUM OF NEW ZEALAND, ALB030841107

ABOVE | A convoy of trucks burns in the desert after an air attack.

NATIONAL ARMY MUSEUM, 2014-13-28

| CHAPTER 5 |

The War in North Africa

With vital strategic interests at stake for both countries, it was inevitable that Italy and Britain would clash in North Africa. For Italian fascist dictator Benito Mussolini, known as Il Duce, the leader who had seized power in 1922, the war offered a chance to make his nation great again. On 10 June 1940, convinced Nazi Germany was winning, Mussolini declared war on Britain and France as Germany's ally. He had two key aims: to turn the Mediterranean into an Italian lake, and to conquer an empire that included North Africa and Greece, thereby transforming Italy into a major world power. Britain, which had established a protectorate in Egypt in order to control the vital Suez Canal, regarded this area as 'the fulcrum of the British Empire'.[1] Britain had also had a presence in the Mediterranean at Malta and Gibraltar since the Napoleonic Wars and was determined to remain there. The region was regarded as second only to the defence of Britain itself.

Having been expelled from the European mainland after a major defeat in France and Belgium in mid-1940, Britain could not face the German Army in Europe alone. Only in North Africa and the Mediterranean, facing a weaker opponent, could Britain commit all three military services to fight. Accordingly, Egypt became the largest British military base outside Britain, built up from Commonwealth forces, including Australian and New Zealand formations.[2] North Africa would be the location of New Zealand's longest ground campaign of the Second World War.

Nazi Germany, with its primary focus on continental Europe, did not have much interest in North Africa or the Mediterranean. Drawn into this theatre of war by Italian ineptitude, it initially committed very few resources to it. However, after General Erwin Rommel's relatively easy victories they began to take the region more seriously. Even then, however, they never put enough manpower and materiel into North Africa until it was too late.

Britain's diversion of valuable assets and forces to Greece in March 1941 coincided with Rommel's arrival in North Africa. During the formation of his Deutsches Afrika Korps (DAK), he was ordered to take things quietly. Not one to follow orders he disliked, Rommel attacked instead and achieved a stunning success. His advance pushed the British out of their newly won positions in Libya, except for the besieged port of Tobruk, and the DAK captured Britain's best field general and army commander, Richard O'Connor. In May and June 1941 two attempts to relieve Tobruk were dismal failures. They cost the British theatre commander General Archibald Wavell his job. His replacement, General Claude Auchinleck from the Indian Army, was ordered to prepare a major offensive to drive Rommel and the Axis forces out of Eastern Libya (Cyrenaica) and relieve Tobruk. This was the aim of Operation 'Crusader' launched on 18 November 1941.

New Zealand's first military engagement in North Africa, 'Crusader' was also the division's first success of the war and its most costly battle. The division fought as part of the Eighth Army, as the British land forces in Egypt had been renamed before 'Crusader', and did so for the rest of the North African campaign. The 'Crusader' offensive was a confusing, fiercely fought series of battles that produced mixed results. Rommel won the initial armoured clash, but he could not overcome the infantry relieving Tobruk. His 'drive to the wire' snatched defeat from victory, exacerbating his losses and his precarious supply position. In early December he broke off the battle and fell back to Gazala and later to El Agheila. Although the British had won, there was no feeling of elation or confidence in the Eighth Army's leadership, equipment or doctrine, and the 'victory' came at the high price of some 18,000 casualties. Thirty per cent of those, some 4620, were New Zealanders, a number greater than the casualties of the Greek and Cretan disasters combined.[3]

AFTERWARDS THE DIVISION was sent to Syria, where it remained on garrison duty for the first five months of 1942. While it was there the Eighth Army suffered its worst defeat of the war in the Battle of Gazala. When Rommel launched an offensive on 27 May 1942 the British knew the attack was coming, and they enjoyed a three-to-one superiority in tanks, and a three-to-two superiority in guns. But the response to the German assault was poorly led. Outmanoeuvred and outfought, the Eighth Army fell back in disarray and did not stop until it reached the El Alamein position, less than 100 kilometres from Alexandria.

Both 2 New Zealand and 9 Australian Divisions, rushed down to Egypt from Syria, were alarmed by the panic they found. They arrived just after Tobruk was captured on 21 June with its entire garrison of 35,000, more men than Rommel had in the DAK. An elated Hitler immediately promoted Rommel to field marshal. On 28 June at Minqar Qaim, the New Zealand Division narrowly escaped capture after it was abandoned by the Eighth Army and surrounded by the German

RIGHT | Senior New Zealand officers confer during the 'Crusader' battle in November 1941. They are, from left, Brigadier William Gentry, unknown (obscured), Brigadier James Hargest, Brigadier Reginald Miles, General Bernard Freyberg and Lieutenant-Colonel Maxwell (far right).

NATIONAL ARMY MUSEUM, 2004-267-3

LEFT | Charles Upham VC (right) and Tom White in silhouette. Second Lieutenant White was captured on 13 December 1941 and, like Upham, spent most of the war as a prisoner in Germany.

NATIONAL ARMY MUSEUM, 2015-40-19

RIGHT | A near miss for 2 New Zealand Division's transports under accurate and heavy artillery fire.

NATIONAL ARMY MUSEUM, 2006-174

ABOVE | The battle headquarters of 20 Battalion in Libya, November 1941.

NATIONAL ARMY MUSEUM, 2015-40-19

BELOW | British Churchill tanks going into action during the Second Battle of El Alamein. Although fast, they were lightly armoured and their 2-pounder gun lacked range and hitting power.

NATIONAL ARMY MUSEUM, 2006-174

ABOVE | New Zealand gunners digging slit trenches at Baggush in the Western Desert.

NATIONAL ARMY MUSEUM, 1987-1930-2

21 Panzer Division. A daring night breakout at bayonet point, which inflicted a severe check on the DAK, was 'the New Zealand Division's most dramatic moment of the whole campaign'. This was the 'perhaps the most savage attack ever carried out by New Zealand troops' and led to recriminations by the Germans and German radio, labelling the New Zealanders 'Freyberg's butchers'.[4]

With the Eighth Army in disarray, Auchinleck took over command on 25 June and made the decision to fall back to El Alamein. Here the Allies would make a determined effort to halt Rommel's advance. The position was well suited for defence. With the Mediterranean Sea in the north and a sea of soft sand, the Qattara Depression, to the south, it could not be bypassed or outflanked. On 1 July 1942 Rommel, down to just 40 tanks and with his soldiers exhausted and existing solely on captured supplies, crashed into the British defences at El Alamein. While there was panic in Alexandria, from which the British fleet sailed away, and in Cairo the two opposing armies slugged it out in three weeks of sporadic fighting now known as the First Battle of El Alamein. Both the New Zealanders and Australians were heavily involved in this fighting, and had some initial success, but then suffered a series of disasters through poor command and coordination. On 15 July, in a night attack on Ruweisat Ridge, New Zealand suffered its worst disaster of the war. Another disaster a week later at El Mreir was almost as bad. Over 1500 men were made prisoners of war in these two actions.[5]

First Alamein was, however, decisive. The Axis forces could not break through and Egypt was saved. Despite this, Churchill had lost confidence in much of the Eighth Army's leadership. Auchinleck and several other generals were replaced. Most soldiers were pleased to see Auchinleck go; this was especially true for the New Zealanders, who had suffered so many losses under his command. Auchinleck's place was taken by General Harold Alexander, with command of the Eighth Army eventually going to the unpopular and inexperienced Lieutenant-General Bernard Montgomery.

THERE WAS A BRIEF stalemate at El Alamein as both sides considered their next moves. Rommel, despite facing huge supply difficulties, knew that time was against him and that he had to strike first. The Italo-German forces attacked on 30 August in what was to become the Battle of Alam Halfa. The Eighth Army, under Montgomery, knew the attack was coming and was well prepared. Rommel's forces were defeated by a combination of dense minefields, a massive artillery concentration and the efforts of the Desert Air Force, commanded by New Zealander Air Vice-Marshal

Arthur 'Mary' Coningham. However, an infantry counterattack launched on 3 September involving the New Zealanders was a notable failure. Despite this, Alam Halfa was a serious blow for Rommel. Morale in the Eighth Army rose and Montgomery, his reputation enhanced, set about preparing an offensive of his own: the Second Battle of El Alamein.

This decisive action, which lasted from 23 October to 4 November 1942, was a turning point in both the North Africa campaign and the war in general. The attack opened with a 900-gun barrage at 9.40 p.m. on 23 October, followed at 10 p.m. by the departure of infantry from five divisions — New Zealanders, South Africans, Australians, Indians and British, some behind a creeping barrage. This attack was only partially successful and soon lost momentum. Montgomery, demonstrating considerable tactical skill, switched the direction of the attack. He used 9 Australian Division in the north to fix the enemy through its 'crumbling' operations, Montgomery's term for wearing down the enemy's defensive positions, while 2 New Zealand Division organised and commanded a new thrust in the south. Operation 'Supercharge', as it was known, succeeded in breaking the resistance of the Italians and Germans. The Axis forces never recovered from this devastating defeat. The Germans lost 10,000 prisoners and the Italians 20,000; also lost was a total of 450 tanks and 1000 guns.[6]

AFTER EL ALAMEIN, the Eighth Army's pursuit of Rommel and his Afrika Korps was badly coordinated and too cautious. Three attempts to outflank and capture the DAK — all involving 2 New Zealand Division — failed, allowing Rommel to fall back to the Mareth Line on the Tunisian–Libyan border, where he awaited the opportunity to strike again. Rommel struck first against the US II Corps in Tunisia at Kasserine Pass, savagely checking the inadequately led and trained Americans. This was a serious wake-up call for the United States Army, 'of importance in setting [it] on the path to becoming a truly effective military organization'.[7]

Thinking it was vulnerable, Rommel then attacked the Eighth Army at Medenine, some 435 kilometres north of Tripoli, on 6 March 1943. Forewarned by Ultra intelligence and reinforced with armour and anti-tank guns, Montgomery again soundly defeated the Germans. Medenine, Rommel's last battle in North Africa, was an unmitigated disaster. He attacked with 124 tanks and lost 52 of them. Five were destroyed by the New Zealanders. No British tanks were lost.[8]

LEFT | A destroyed Italian tank burns in the desert.

NATIONAL ARMY MUSEUM, 2006-174

ABOVE | A tank battle is photographed as it happens.

NATIONAL ARMY MUSEUM, 1992-395-4

RIGHT | A badly damaged Italian tank, its turret blown off, at Sidi Rezegh in November 1941. Like the British Crusaders, the Italian tanks were lightly armoured and underpowered.

NATIONAL ARMY MUSEUM, 1992-1161-1

RIGHT | Infantrymen advance into the smoke and dust of a desert action.

NATIONAL ARMY MUSEUM, 1992-395-4

BELOW | Soldiers dig in on a ridgeline in the desert.

NATIONAL ARMY MUSEUM, 2004-420

LEFT | Charles Bennett, driver of the Te Aroha Canteen Truck for 28 (Maori) Battalion, in a dugout at Sidi Rezegh.

NATIONAL ARMY MUSEUM, 2012-2-1

THERE WAS STILL HEAVY fighting after Rommel left North Africa, as Hitler sent thousands of German soldiers and a new army commander, Generaloberst Hans-Jürgen von Arnim, to try to hold on to North Africa. On 20 March 1943 Montgomery launched an assault on the Mareth Line while the New Zealand Division, boosted to a corps with an additional armoured brigade, made a difficult march to reach Tebaga Gap, a low mountain pass behind the enemy's position. When the frontal assault on the Mareth Line failed, Montgomery again switched the direction of attack, reinforcing Freyberg's corps with a complete armoured division and giving it the main assault role. The renewed attack on 26 March punched through the Mareth Line in a blitzkrieg-style operation with perfect coordination between ground and air forces.

Now the Axis forces in North Africa were confined to northern Tunisia. They were steadily squeezed between the First Army, as part of the force that landed in North Africa, in Operation 'Torch', closing in from the north-west, and the Eighth Army. There was some stiff fighting in Tunisia, but the Axis troops were being starved of supplies. All German convoys ceased by the end of March 1943 and an attempt to maintain an airbridge resulted in the loss of more than 200 transport aircraft.[9] Without fuel, only limited ammunition and their backs to the Mediterranean, there was only one option. On 13 May 1943, the Axis forces in North Africa surrendered: some 275,000 men, including 12 generals, laid down their arms.[10] Italy would surrender to the Allies four months later. For the Germans it was a defeat to match that of Stalingrad, a few months before.

The North African campaign was always a sideshow for the Germans. For most of the fighting there Germany had only three divisions, compared with over 200 on the Eastern Front. For Britain and Italy, though, North Africa was their main effort from 1940 to 1943. The battles of El Alamein, Medenine and Tunisia were part of the major shift in the war known as 'the turn of the tide'.[11] The German string of victories was over; Hitler had finally lost the initiative. The 2nd New Zealand Division had been at the forefront of all the Eighth Army's successes in North Africa, and on the receiving end of several reversals, especially Ruweisat Ridge. But the achievements had been hard won, paid for in 'torment, thirst, hatred and blood'.[12] And the war was far from over.

ABOVE | A 25-pounder gun crew in action at Azizia, Libya.

NATIONAL ARMY MUSEUM, 1994-2205-2

RIGHT | Armoured cars on a reconnaissance patrol in the desert.

NATIONAL ARMY MUSEUM, 2006-174

LEFT | Soldiers watch as the Libyan port city of Benghazi is bombed.

NATIONAL ARMY MUSEUM, 1994-2827

BELOW | 2 New Zealand Division transports under shellfire at Sidi Rezegh in November 1941.

NATIONAL ARMY MUSEUM, 2004-267-2

RIGHT | New Zealand trucks ablaze at Sidi Rezegh.

NATIONAL ARMY MUSEUM, 2004-267-2

RIGHT | Chaos as the battle rages around Tobruk in November 1941.

NATIONAL ARMY MUSEUM, 2004-267-2

LEFT | A 25-pounder at full recoil during the 'Crusader' battle.

NATIONAL ARMY MUSEUM, 2004-267-3

BELOW | Medical staff attend to the wounded at a main dressing station (MDS) in the desert.

NATIONAL ARMY MUSEUM, 1990-1206

RIGHT | A church service in the sand at Tobruk.

NATIONAL ARMY MUSEUM, 1991-770

ABOVE | New Zealand wounded on stretchers under the relentless desert sun.

NATIONAL ARMY MUSEUM, 1990-1206

RIGHT | General Freyberg, left, and his staff observe the battle and do not like what they see.

NATIONAL ARMY MUSEUM, 2004-267-3

LEFT | Freyberg (centre) cuts a lonely figure as he watches the action at Sidi Rezegh.

NATIONAL ARMY MUSEUM, 2004-267-3

ABOVE | As artillery shells burst around him, Freyberg speaks on the radio at Sidi Rezegh.

NATIONAL ARMY MUSEUM, 2004-267-3

RIGHT | A 15-year-old German prisoner of war captured by the New Zealanders at Sidi Rezgh in November 1941.

NATIONAL ARMY MUSEUM, 1992-1162

RIGHT | Senior Luftwaffe officers greet German aircrew arriving in Africa.

NATIONAL ARMY MUSEUM, 1993-1313-1

BELOW | Sergeant Percy Cross from Wellington, on the right, and a mate after the ordeal at Sidi Rezegh in November 1941. Cross was a member of 8 Field Company, New Zealand Engineers.

NATIONAL ARMY MUSEUM, 1990-287-1

ABOVE | Brigadier Howard Kippenberger returns to Belhamed, Libya, where his 20 Battalion was overrun in November 1941.

NATIONAL ARMY MUSEUM, 2015-40-19

LEFT | New Zealand gunners enjoying their 1941 Christmas dinner, eaten out of dixies.

NATIONAL ARMY MUSEUM, 2017-640-2

LEFT | Players compete for the ball in a lineout during a rugby match in the desert.

STEPHEN PARSONS COLLECTION

ABOVE | Washing day in Syria for members of this artillery unit. A welcome break away from the Western Desert, the time in Syria allowed the New Zealand Division to train and revise its procedures after the 'Crusader' campaign.

NATIONAL ARMY MUSEUM, 1990-144

LEFT | General Claude Auchinleck, second from left, talks with one of his brigadiers. Although he stopped Rommel in the First Alamein battle, Auchinleck was not a great success as an army or theatre commander. Freyberg and other senior New Zealand commanders were pleased to see him replaced by General Bernard Montgomery.

NATIONAL ARMY MUSEUM, 1993-1313-1

RIGHT | New Zealand officers confer in the Kaponga Box (a fortified defensive position) near the Qattara Depression. Being locked up in a defensive position like this was anathema to Freyberg and his men.

NATIONAL ARMY MUSEUM, 1992-1162

LEFT | A desert column under air attack. A bomb has exploded near the vehicles.

STEPHEN PARSONS COLLECTION

ABOVE | An anti-aircraft gun crew of 14 Light Anti-Aircraft Regiment, New Zealand Artillery, pose with a captured Italian Breda anti-aircraft gun.

NATIONAL ARMY MUSEUM, 2017-640-2

LEFT | Digging defensive positions with a drill and shovel. The drill made it much easier, but it was still hard work.

NATIONAL ARMY MUSEUM, 2005-587-1

LEFT | 'One we got': the anti-aircraft gunners have scored a hit.

NATIONAL ARMY MUSEUM, 2017-640-1

BELOW | Gunner Eric Groves from Papatoetoe in his dugout.

NATIONAL ARMY MUSEUM, 1987-1930-3

BELOW | 'Two Gunner mates', it says on the back of this photograph of Albert Wilson and Joseph Summers outside their dugout. Twenty-seven-year-old Summers was killed in action during the First Battle of El Alamein on 5 July 1942.

NATIONAL ARMY MUSEUM, 1987-1930-3

LEFT | A machine gun position in the desert has attracted enemy artillery fire.

NATIONAL ARMY MUSEUM, 2006-174

BELOW | Artillery shells land near a unit's field headquarters.

NATIONAL ARMY MUSEUM, 1992-395-4

RIGHT | An anti-tank gun in action.

NATIONAL ARMY MUSEUM, 2006-174

ABOVE | The wreckage of a plane and the graves of the passengers and crew who died when it was shot down on 7 August 1942. One of those buried was Lieutenant General W. H. E. (Strafer) Gott, who had just been appointed Eighth Army commander. Gott's death paved the way for Montgomery's appointment.

NATIONAL ARMY MUSEUM, 2006-174

ABOVE | Two hours after being captured on 4 September 1942, Brigadier George Clifton, left, meets Field Marshal Rommel, centre with his arms behind his back. Rommel opened the conversation through his interpreter (Major Burckhardt, second on left) with a long harangue about the New Zealanders' 'gangster fighting methods' at Minquar Quain. He seemed, however, to accept Clifton's defence, and also asked why the New Zealanders were fighting in North Africa. This photograph was taken by a German soldier who was killed in action a month later. His camera was souvenired and the film developed. Rommel was widely admired in the Eighth Army. Many soldiers and nurses had his photograph in their albums.

NATIONAL ARMY MUSEUM, 1992-1084-5

LEFT | Italian prisoners of war, taken at Halfaya Pass in November 1941, marching into captivity with a minimal escort.

NATIONAL ARMY MUSEUM, 1992-395-4

BELOW | Heavy going on manoeuvres for 27 Machine Gun Battalion.

NATIONAL ARMY MUSEUM, 1992-1084-3

RIGHT | This 27 Machine Gun Battalion position in the desert was likely to be a training exercise since it was unwise to stand silhouetted on a ridgeline.

NATIONAL ARMY MUSEUM, 1993-1311

LEFT | A soldier teases his friends with an unexploded Italian bomb.

NATIONAL ARMY MUSEUM, 1993-1311

BELOW | New Zealand transports at Sidi Rezegh under artillery fire.

NATIONAL ARMY MUSEUM, 1992-1162

ABOVE | A post box in the field. As with the First World War, a regular mail service was a vital link for all service personnel and essential to maintaining morale.

NATIONAL ARMY MUSEUM, 1990-1206

RIGHT | Keeping boredom at bay. Eric Jeffrey (left) and Jock Clouston put their musical instruments to good use.

NATIONAL ARMY MUSEUM, 2006-40

ABOVE | That vital lifeline to New Zealand, letters from home.

NATIONAL ARMY MUSEUM, 2004-420

RIGHT | New Zealand soldiers with an unexploded 250-pound bomb.

NATIONAL ARMY MUSEUM, 1990-144

LEFT | The crew of a 25-pounder eat their meal in the desert.

NATIONAL ARMY MUSEUM, 1991-980

RIGHT | This conference in the desert includes General Freyberg, second from left, and, on the right, British foreign secretary Anthony Eden.

NATIONAL ARMY MUSEUM, 1992-395-4

ABOVE | Stopping to brew up.

NATIONAL ARMY MUSEUM, 2006-174

RIGHT | New Zealand gunners examine a destroyed German Mark III medium tank.

NATIONAL ARMY MUSEUM, 1991-980

ABOVE | On the right of this group of 28 (Maori) Battalion officers is 2nd Lieutenant Te Moananui-a-Kiwa Ngarimu, who would be awarded a posthumous Victoria Cross for his exceptional gallantry at Tebaga Gap in March 1943.

NATIONAL ARMY MUSEUM, 2012-2-1

LEFT | The newly commissioned Keith Elliott VC with New Zealand nurse Sister Violet Wallace.

NATIONAL ARMY MUSEUM, 1993-2420-2

ABOVE | A silhouette of soldiers moving forward at dusk.

NATIONAL ARMY MUSEUM, 1992-395-4

RIGHT | Laying a pipeline in the desert was hot, dirty and back-breaking work.

NATIONAL ARMY MUSEUM, 1991-770

LEFT | Flies were a constant irritation. A soldier checks a fly trap outside a New Zealand hospital.

NATIONAL ARMY MUSEUM, 2000-594-1

BELOW | Wearing fly veils helped.

NATIONAL ARMY MUSEUM, 1992-1084-8

ABOVE | Having a wash — water was always in short supply in North Africa.

NATIONAL ARMY MUSEUM, 1992-395-4

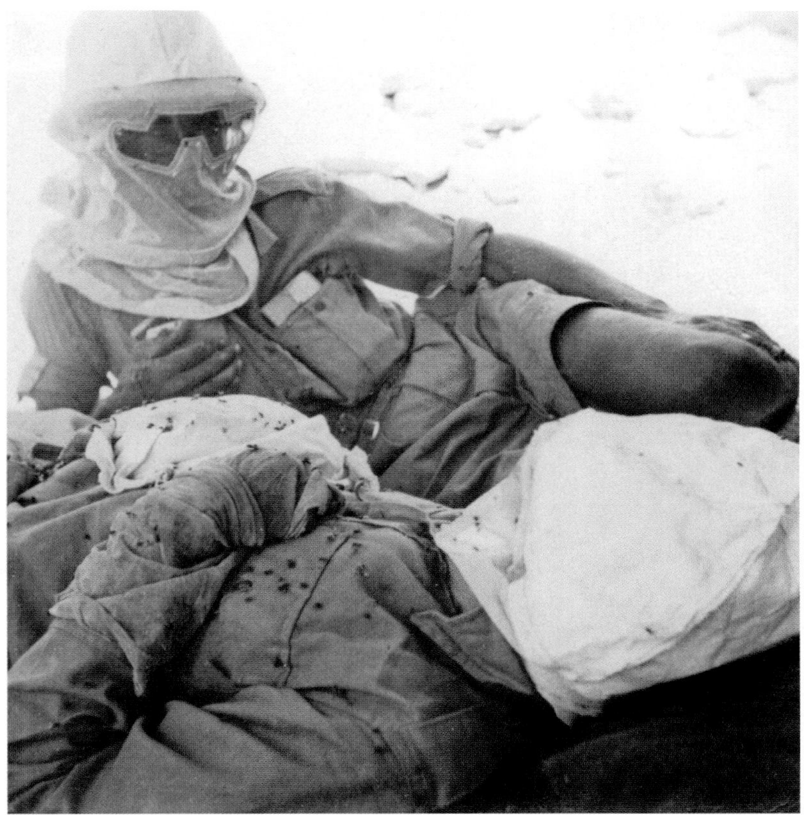

LEFT | This photograph appeared in the albums of many New Zealand soldiers, who understood how essential it was to make a cuppa whenever possible.

NATIONAL ARMY MUSEUM, 1992-395-4

BELOW | 'A watched pot': Harry Stephenson brewing the ubiquitous desert tea.

NATIONAL ARMY MUSEUM, 1992-1084-9

ABOVE | As shown by this, and the following group of photographs, there are a large number of German photographs in New Zealand collections. Here, German soldiers take the opportunity to clean their personal weapons during a lull at an artillery position at Bardia.

NATIONAL ARMY MUSEUM, 2006-49

LEFT | A confident-looking group of German soldiers.

NATIONAL ARMY MUSEUM, 1992-1084-9

LEFT | Three German soldiers pose by a field gun.

NATIONAL ARMY MUSEUM, 1992-1084-9

BELOW | This German anti-tank gun firing on the Eighth Army appears to have scored several hits.

NATIONAL ARMY MUSEUM, L15-2010-1

BELOW | A German half-track reconnaissance unit in action.

NATIONAL ARMY MUSEUM, L15-2010-1

RIGHT | German soldiers on lookout.

NATIONAL ARMY MUSEUM, L15-2010-1

BELOW | The weary crew of a German half-truck taking a break in the desert.

NATIONAL ARMY MUSEUM, L15-2010-1

LEFT | 'Four good Jerries' reads the annotation to this photograph. German graves under a palm tree far from home.

NATIONAL ARMY MUSEUM, 1992-395-4

BELOW | David Parsons of 27 Machine Gun Battalion poses in front of an American-manufactured Grant tank, the first the Eighth Army used that could match German tanks in terms of armour and firepower. However, its gun position in a sponson (the attached compartment on the side of a tank) rather than a revolving turret was a serious limitation.

STEPHEN PARSONS COLLECTION

BELOW | New Zealand soldiers ride on a Quad (field artillery tractor) flying an unusual flag.

NATIONAL ARMY MUSEUM, 1994-2205-2

ABOVE | An anti-aircraft gun crew of 14 Light Anti-Aircraft Regiment with some of their captured German souvenirs.

NATIONAL ARMY MUSEUM, 1987-1930-2

RIGHT | A low-flying 'Wimpy' (Vickers Wellington bomber) over the desert. Air support was essential to the outcome in North Africa.

NATIONAL ARMY MUSEUM, 1992-395-4

ABOVE | A member of the New Zealand Engineers relaxes with a cigarette and a book.

NATIONAL ARMY MUSEUM, 1993-1312

BELOW | Members of 28 (Maori) Battalion preparing a hangi for Christmas Day 1942 at Nofilia, Libya.

NATIONAL ARMY MUSEUM, 2012-2-1

RIGHT | Lieutenant George Katene of 28 (Maori) Battalion meets up with his sister Wikitoria Katene in North Africa. Wikitoria (Wiki) Katene was the first Māori woman to serve as a nurse in the Second World War. Her brother George, who had been awarded the Military Medal for his gallantry in the Greek campaign, was later killed in action in the fighting around Orsogna, Italy, on 7 December 1943.

NATIONAL ARMY MUSEUM, 2012-2-1

ABOVE | Wrecked German planes held a fascination for New Zealand soldiers. Members of the Kiwi Concert Party pose with a downed Messerschmitt 109 (Bf 109) fighter.

NATIONAL ARMY MUSEUM, 2003-426

LEFT | Recovering after a few too many beers in the desert.

NATIONAL ARMY MUSEUM, 1992-47

RIGHT | Searchlights and anti-aircraft fire light up the sky during an air raid.

NATIONAL ARMY MUSEUM, 1994-2827

LEFT | 'Don' poses with his war trophies.

NATIONAL ARMY MUSEUM, 2004-420

BELOW | A German cemetery attests to the hard fighting in North Africa.

NATIONAL ARMY MUSEUM, 1992-1160

ABOVE | A truck ploughs through the desert dust.

NATIONAL ARMY MUSEUM, 1993-1312

RIGHT | The X-ray truck of a New Zealand Field Ambulance has suffered some damage.

NATIONAL ARMY MUSEUM, 1988-2090

LEFT | The dispensary of No. 3 New Zealand General Hospital.

NATIONAL ARMY MUSEUM, 2000-594-1

LEFT | New Zealand nurses wrapped up against the desert cold after coming off night shift. At the far right, kneeling, is Violet Wallace.

NATIONAL ARMY MUSEUM, 1993-2420-2

RIGHT | Nurses from No. 2 New Zealand General Hospital at Gerawla, Egypt, take time out in November 1941.

NATIONAL ARMY MUSEUM, 1993-2420-2

LEFT | Violet Wallace (right) and a fellow nurse, most likely at Gerawla.

NATIONAL ARMY MUSEUM, 1993-2420-2

LEFT | At Mersa Matruh a New Zealand nurse points to one of the prominent dangers in the region.

NATIONAL ARMY MUSEUM, 1993-2420-2

BELOW | Nancy White, from Feilding, attends to a patient with a leg injury.

NATIONAL ARMY MUSEUM, 2000-954-1

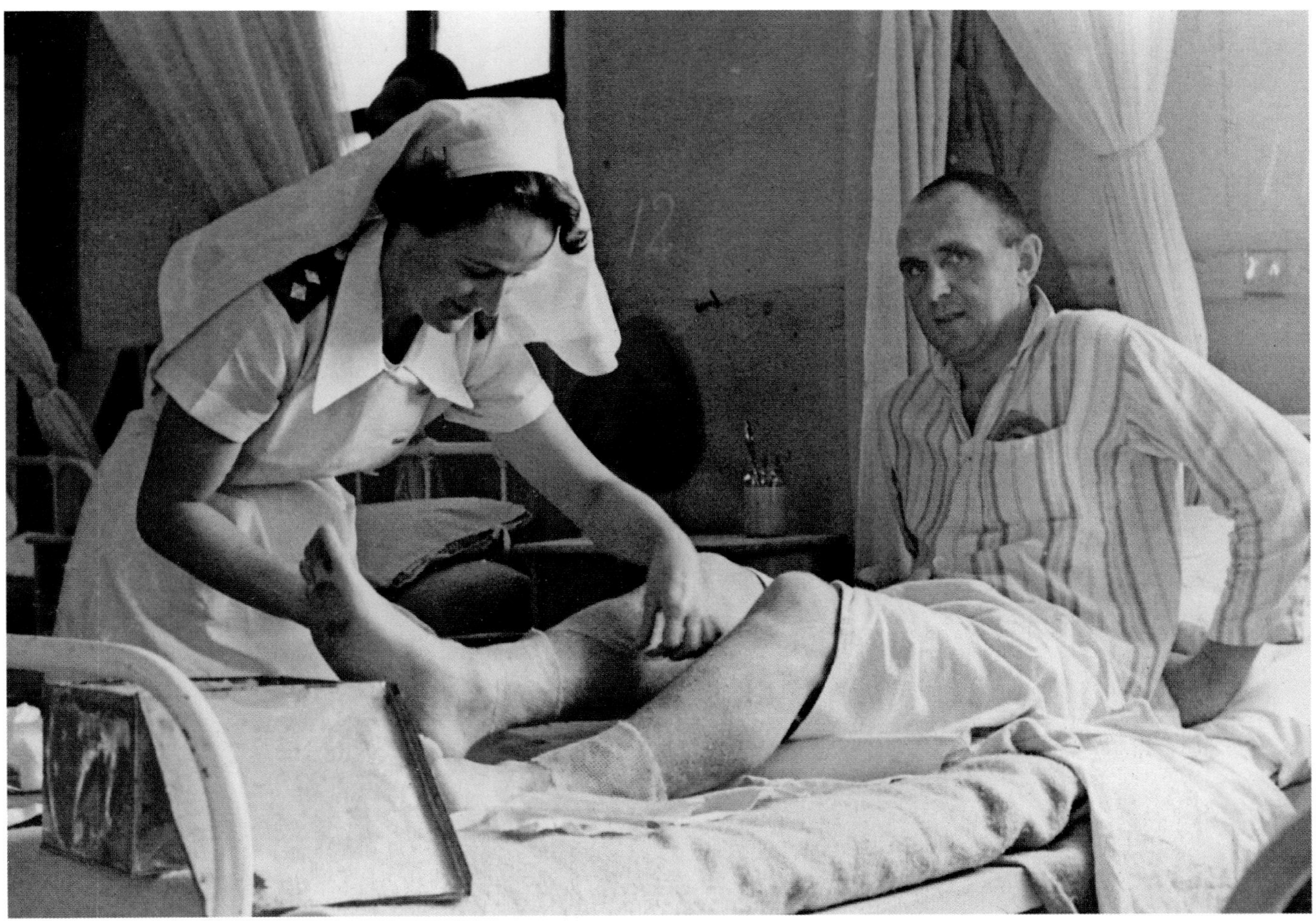

RIGHT | Tough work in tough conditions — laying railway tracks in the desert.

NATIONAL ARMY MUSEUM, 1990-389-8

ABOVE | Lining up the tracks of the desert railway.

NATIONAL ARMY MUSEUM, 1993-1312

RIGHT | A captured Deutsches Afrika Korps truck in the hands of the New Zealand Railway Company.

NATIONAL ARMY MUSEUM, 1990-389-3

ABOVE | Three soldiers of 16 Railway Company pose in front of the minefield near Minqar Qaim, Syria.

NATIONAL ARMY MUSEUM, 1990-389-4

ABOVE | Another image of the most photographed senior officer in North Africa, but here Rommel looks thin, emaciated and unwell. His health deteriorated from August 1942.

ENGINEER CORPS MEMORIAL MUSEUM, WALLACE 100

BELOW | Generals Sir Oliver Leese, Freyberg and Montgomery enjoying a performance by the Kiwi Concert Party.

NATIONAL ARMY MUSEUM, 1988-1319

ABOVE | Before the Second Battle of El Alamein, the Eighth Army received new electric Polish mine detectors and flail tanks, but there were not enough of them and the old-fashioned method of prodding the ground with a bayonet had to be used.

NATIONAL ARMY MUSEUM, 2006-174

ABOVE | A soldier manning the anti-aircraft gun keeps an eye on the Kiwi Concert Party's performance.

NATIONAL ARMY MUSEUM, 1988-1319

LEFT | A New Zealand sapper using his bayonet to clear a German minefield has just found and defused a German Teller mine. Success in the Second Battle of El Alamein depended on the work of the Engineers.

ENGINEER CORPS MEMORIAL MUSEUM, RNZE HISTORY

RIGHT | Two sappers defuse German anti-personnel mines. At El Alamein, more than 500,000 mines were laid to prevent the Eighth Army from making a breakthrough.

NATIONAL ARMY MUSEUM, 2006-174

ABOVE | Arthur Rayner of 36 (NZ) Survey Battery, New Zealand Artillery, outside a camouflaged spotting tent at El Alamein in October 1942.

NATIONAL ARMY MUSEUM, 2007-989-1

RIGHT | A New Zealand corporal locates and defuses a German anti-personnel mine. Both sides used mines extensively in the desert.

NATIONAL ARMY MUSEUM, 1992-395-4

LEFT | The battle begins. The 900-gun barrage signals the start of the Second Battle of El Alamein on 23 October 1942. This photograph was taken from the New Zealand main dressing station (MDS).

NATIONAL ARMY MUSEUM, 1990-1206

LEFT | A New Zealand 25-pounder in action at El Alamein on the night of 23 October 1942.

NATIONAL ARMY MUSEUM, L15-2010-1

ABOVE | Infantry soldiers advance in the open.

NATIONAL ARMY MUSEUM, 1992-395-4

BELOW | Soldiers watch as a shell explodes near a transport vehicle at El Alamein.

NATIONAL ARMY MUSEUM, 2004-267-3

ABOVE | Eighth Army soldiers watch an artillery barrage fired at night.

NATIONAL ARMY MUSEUM, 2006-174

BELOW | The first Sherman tanks supplied to the Eighth Army advance in desert formation.

NATIONAL ARMY MUSEUM, 2006-174

RIGHT | A soldier seeks shelter as artillery rounds land near a minefield.

NATIONAL ARMY MUSEUM, 2004-267-3

BELOW | General Bernard Montgomery, the victor of El Alamein, climbing down from the Grant tank he used as his battle headquarters.

NATIONAL ARMY MUSEUM, 2006-174

ABOVE | Infantry advance with fixed bayonets.

NATIONAL ARMY MUSEUM, 2006-174

LEFT | A tank crew surrendering to advancing infantry. This is likely to be a posed photograph: tanks were deadly infantry killers, impervious to rifle fire and bayonet.

NATIONAL ARMY MUSEUM, 2006-174

RIGHT | A British Kittyhawk fighter firing its guns at night.

NATIONAL ARMY MUSEUM, 1994-2827

LEFT | Padre Edward (Ted) Forsman conducting a triple burial service near a New Zealand main dressing station at El Alamein on 24 October 1942.

NATIONAL ARMY MUSEUM, 1990-1206

BELOW | Helmets strewn on the battlefield attest to the number of casualties suffered.

NATIONAL ARMY MUSEUM, 2006-174

RIGHT | German dead litter the battlefield after the El Alamein battle.

NATIONAL ARMY MUSEUM, 2006-174

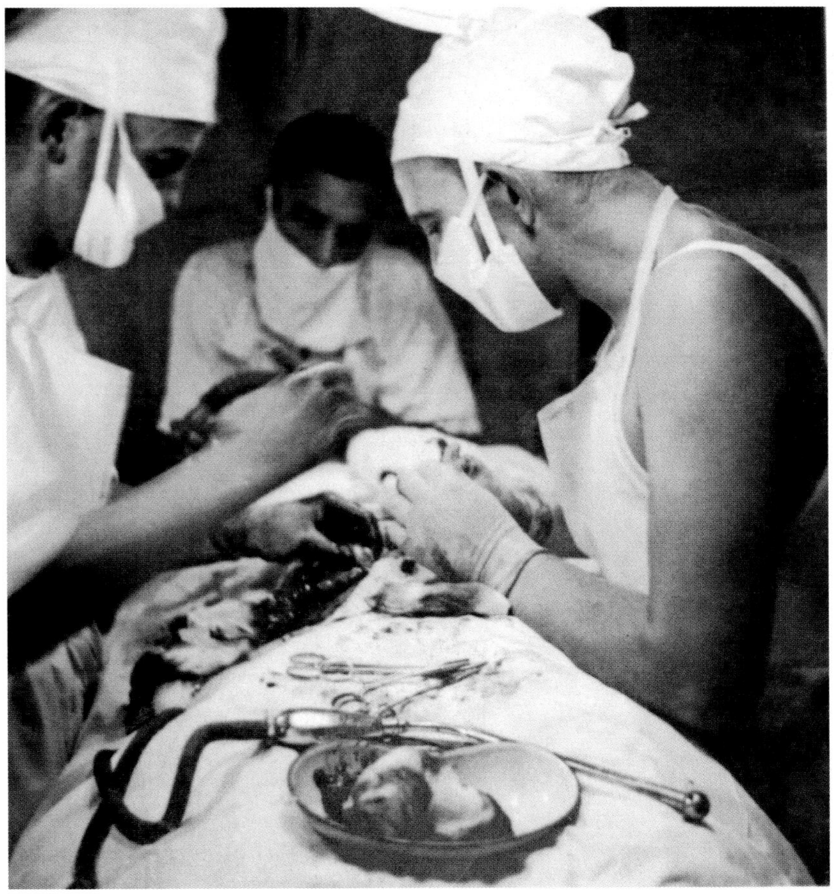

ABOVE | German prisoners of war captured after the Second Battle of El Alamein.

NATIONAL ARMY MUSEUM, 1990-1206

RIGHT | Major S. L. Wilson performs emergency surgery at a main dressing station.

NATIONAL ARMY MUSEUM, 1990-1206

ABOVE | Soldiers locate friends' or relatives' graves at the El Alamein cemetery, which contains 7240 Commonwealth men, 815 of whom are unidentified. More than 1100 New Zealanders are buried or commemorated here.

NATIONAL ARMY MUSEUM, 1990-144

RIGHT | New Zealand soldiers relax at the El Alamein railway station.

NATIONAL ARMY MUSEUM, 2006-174

ABOVE | A truck of the Long Range Desert Group (LRDG) advances cautiously over a sand dune. The LRDG was an elite unit, one of the early special forces like commandos. Many New Zealanders joined it during the war in North Africa.

NATIONAL ARMY MUSEUM, 1998-444

ABOVE | Some men of the LRDG with their truck. The Silver Fern on his jersey indicates the nationality of one member.

NATIONAL ARMY MUSEUM, 2014-13-28

RIGHT | An LRDG crew with their trucks.

NATIONAL ARMY MUSEUM, 1998-444

ABOVE | As this corporal demonstrates, LRDG soldiers were heavily armed with automatic weapons.

NATIONAL ARMY MUSEUM, 1998-444

RIGHT | An LRDG soldier on lookout.

NATIONAL ARMY MUSEUM, 1998-444

LEFT | LRDG trucks transverse a slope in French West Africa.

NATIONAL ARMY MUSEUM, 1998-444

ABOVE | Eighth Army tanks approaching Mersa Matruh after the October El Alamein battle.

NATIONAL ARMY MUSEUM, 1990-1206

LEFT | Smiles all round from a group of New Zealand gunners at Nofilia in Libya in 1942. The war is finally going well and Christmas dinner is imminent.

NATIONAL ARMY MUSEUM, 2017-640-2

ABOVE | New Zealand soldiers of 6 RMT awaiting their Christmas beer issue at Nofilia in 1942. The soldier at the arrow marked 'Me' is Ronald Wulff.

NATIONAL ARMY MUSEUM, 1999-693

RIGHT | German planes destroyed after their airfield was overrun.

NATIONAL ARMY MUSEUM, 1994-2827

BELOW | Curious bystanders watch as the Eighth Army enters Tripoli on 23 January 1943. The armoured cars of 11 Hussars, 7 Armoured Division, was the first unit to enter the city. These are Valentine tanks of 40 Royal Tank Regiment, 23 Armoured Brigade.

NATIONAL ARMY MUSEUM, 1990-1206

LEFT | The 28 (Maori) Battalion march past in the Tripoli parade.

NATIONAL ARMY MUSEUM, 2012-2-1

RIGHT | Soldiers of 28 (Maori) Battalion compete in a tug-of-war event at Tripoli.

NATIONAL ARMY MUSEUM, 2012-2-1

BELOW | A 25-pounder in action at Takrouna.

NATIONAL ARMY MUSEUM, 1991-80

ABOVE | Maori Battalion soldiers examine a German Mark III tank destroyed during the Battle of Medenine in Tunisia in March 1943.

NATIONAL ARMY MUSEUM, 2012-2-1

ABOVE | New Zealanders standing on Takrouna look down on the Enfidaville Plain below. Takrouna was an important feature, and 2 New Zealand Division suffered considerable casualties in taking it. Its capture, primarily by the soldiers of 28 (Maori) Battalion, was described by one British general as 'the most gallant feat of arms I witnessed in the course of the war'.

NATIONAL ARMY MUSEUM, 1994-2827

RIGHT | A German heavy gun captured in Tunisia.

NATIONAL ARMY MUSEUM, 1990-144

RIGHT | The graves of Māori soldiers at Takrouna.

NATIONAL ARMY MUSEUM, 2012-2-1

BELOW | E Troop of the New Zealand Artillery return to Maadi Camp from Tunisia at the end of a long, hard campaign.

NATIONAL ARMY MUSEUM, 1991-980

ABOVE | The converted battlecruiser and aircraft carrier HMS *Courageous* after being hit by two torpedoes fired by German U-boat *U29* in the first month of the war. *Courageous*, the first British warship sunk by the enemy during the war, capsized and sank within 20 minutes, with the loss of 519 of her crew. All the personnel on *U29* were awarded the Iron Cross.

AIR FORCE MUSEUM OF NEW ZEALAND, ALB8737528025

| CHAPTER 6 |

The War at Sea

Technically there was no New Zealand navy in 1939. There was instead a New Zealand Division of the Royal Navy — a small flotilla of an imperial navy with worldwide commitments. The demands of war, however, rapidly transformed New Zealand's naval forces in terms of size, personnel and function.

In September 1939, the two main vessels of the New Zealand Division passed to control of the British Admiralty, sailing away to link up with other parts of the Royal Navy on operations far from New Zealand's shores. While *Achilles* headed to America and the West Indies Station, *Leander* escorted Tasman convoys and was then stationed in the Indian Ocean. *Achilles* was soon in action against the pocket battleship *Admiral Graf Spee* in the Battle of the River Plate. In an 'action-packed deployment lasting more than a year', *Leander* escorted convoys under air attack, helped to sink an Italian submarine, clashed with Vichy French destroyers off the coast of Syria and, in February 1941, sank the Italian merchant raider *Ramb I*.[1]

On 1 October 1941, the New Zealand Division of the Royal Navy became the Royal New Zealand Navy (RNZN). Although this gave New Zealand the same status as other British dominions, the government had not supported or requested the change, believing it might seem selfish and disloyal in such a time of crisis. All that really changed, though, was that the 'HMS' prefix to New Zealand vessels became 'HMNZS': His Majesty's New Zealand Ship.

All the New Zealanders who joined the navy during the Second World War were volunteers. Most who enlisted did so for 'Hostilities Only'; that is, for the duration of the war. More than 7000 of them served in the Royal Navy. The first batch arrived in mid-1940 and a steady stream followed. Most of these initial volunteers served on minesweepers in the English Channel, protecting British shipping. From 1941, though, New Zealanders served throughout the Royal Navy and in every theatre of the naval war. The worst disaster of the war for New Zealand navy personnel occurred in December 1941 when the light cruiser HMS *Neptune* hit a mine off the coast of Tripoli. All but one of the crew lost their lives. HMS *Neptune* had been earmarked for transfer to the RNZN and 150 of its crew were New Zealanders. They were all killed. Another 40 New Zealanders serving with the Royal Navy were killed and a similar number taken prisoner during Japanese attacks on Hong Kong, Malaya and Singapore. Many New Zealanders took part in the Battle of the Atlantic, which lasted for most of the war, and served on vessels escorting the Arctic convoys of merchant ships that carried supplies to northern Russia. Around 200 New Zealanders volunteered for submarine service. At the time of the D-Day landings in June 1944, more than 4700 New Zealanders were serving with the Royal Navy, which provided most of the naval assets for this renowned amphibious assault on the beaches of Normandy.[2]

A BRANCH OF THE Royal Navy in which New Zealanders excelled was the Fleet Air Arm. During the Second World War air power proved to be the dominant form of warfare, even at sea. The big naval battles of the war like Coral Sea (May 1942), Midway (June 1942) and Leyte Gulf (October 1944) were primarily air clashes. By then aircraft carriers, not large battleships, were the kingpins of a fleet, as the sinking of the *Repulse* and *Prince of Wales* in December 1941 had clearly showed.

The Royal Navy had been slow to absorb the importance of aircraft in naval actions. Although the Air Branch of the Royal Navy opened recruitment to members from the dominions in July 1939, by April 1940 just three New Zealanders had joined. It was only when 30 per cent of the Fleet Air Arm were lost in the Norwegian campaign between early April and early June that an intensive recruitment campaign began, with an appeal to the dominions for help. New Zealand offered 20 candidates a month and sent its first draft in July 1940, with higher numbers in the following three months. In November that year, 18 members of the fifth draft of pilots were captured when their ship *Rangitane* was sunk by the German raiders *Orion* and *Komet*. Following an Admiralty request in October, New Zealand doubled its commitment to 40 personnel a month.[3] In all, 1066 New Zealanders were selected and served in the Fleet Air Arm; 738 qualified as pilots, making up some 10 per cent of the total number.[4] Nineteen New Zealanders commanded Fleet Air Arm squadrons.

In the last nine months of the war, as part of British Pacific Fleet, the Fleet Air Arm played a major role in defeating Japan. A staggering 25 per cent of the personnel involved in these operations were New Zealanders.[5] Some 150 New Zealanders were killed in action flying for the Royal Navy. They included the last combat death of the war: Sub-Lieutenant Thomas McBride, who died on 10 August 1945.[6]

Some prominent New Zealanders served in the Royal Navy. They included Peter Phipps, who became the first New Zealand-born chief of naval staff; Phil Connolly, a minister of defence in the 1950s; and James (Jim) Macdonald, who

ABOVE | British merchant seamen celebrate their arrival in Britain in February 1940. They had been prisoners on the German ship *Altmark* after their own ships had been sunk by the *Graf Spee*. The age range is striking. No New Zealand civilians were more at risk than merchant seamen; two 15-year-old deck boys were the youngest New Zealanders killed in combat in the twentieth century.

NATIONAL MUSEUM OF THE ROYAL NEW ZEALAND NAVY, OC1064

RIGHT | Some of the passengers who had been on the mail steamer *Niagara* when it struck a mine in the Hauraki Gulf in June 1940, about to be taken in tow by motor boats of HMS *Achilles*.

NATIONAL MUSEUM OF THE ROYAL NEW ZEALAND NAVY, OC1244

LEFT | Fleet Air Arm (FAA) pilots at a ground dispersal hut, probably during the Battle of Britain when the RAF was desperately short of fighter pilots. They are named as Jock Anderson, Bill Bowe, Swkewke, Jackie Ramsden, Twysel, Bill McGuiness and Jim Bignell. Some 23 naval pilots served in various Fighter Command squadrons during the battle, while two FAA squadrons provided air cover for Britain's southern dockyards.

NATIONAL MUSEUM OF THE ROYAL NEW ZEALAND NAVY, 2014.32.28.P18.1

LEFT | SS *Rotorua* photographed off the coast of Scotland on 11 December 1940. Moments later, the vessel was sunk by a German submarine.

NATIONAL MUSEUM OF THE ROYAL NEW ZEALAND NAVY, OC1107

ABOVE | Able Seaman Jack True mans the anti-aircraft gun in January 1941.

NATIONAL MUSEUM OF THE ROYAL NEW ZEALAND NAVY, 2006.186.16.P33.1

RIGHT | HMNZS *Moa* at Leith, Scotland, in October 1941 shortly after being fitted out as a minesweeper.

NATIONAL MUSEUM OF THE ROYAL NEW ZEALAND NAVY, OC138

commanded a flotilla of torpedo boats and became the Second World War's most decorated New Zealand sailor. He was awarded the DSO, the DSC with two bars and was Mentioned in Dispatches (MID) twice, all before the age of 23. While commanding the unit, he also managed to invent an effective torpedo sight.[7] Archibald Richardson of the Fleet Air Arm flew a Hellcat dive-bomber against the German battlecruiser *Tirpitz* and was commander of 1840 Naval Air Squadron on *Indefatigable*. After being killed in action, he was recommended for a posthumous VC, only to receive an MID instead.[8] It seems unfair that no award of a Victoria Cross was made to New Zealand navy personnel, although others were recommended for it. During the war, 449 New Zealanders were killed in action while serving with the Royal Navy. Of the 63 made prisoners of war, nine died in captivity.[9]

TO PROTECT NEW ZEALAND from naval threats HMS *Monowai* was fitted out as an armed cruiser to patrol New Zealand waters, and shore batteries were established and garrisoned to protect the country's harbours. It was not enough: New Zealand was isolated and vulnerable. On 13–14 June 1940, the German raider *Orion* laid 228 mines across the approaches to Auckland in the Hauraki Gulf. Five days later the mail steamer *Niagara*, carrying 136 passengers, half of New Zealand's small arms ammunition and £2.5 million in gold bullion, hit one of these mines. No lives were lost but the gold and ammunition were. On 14 May 1941, the minesweeper HMS *Puriri* hit one of the mines and sank. Five seamen were killed, the only New Zealand naval personnel to die in home waters. On 20 August 1940, the *Orion* intercepted the *Turakina* in the Tasman Sea, some 640 kilometres from New Zealand. In the ensuing gun battle some Germans were wounded, but 35 *Turakina* crew were killed and 21 became prisoners of war (though one died of his injuries).[10] This was the first and only engagement in the Tasman Sea and the closest to New Zealand. Other enemy vessels, including a German submarine, operated in New Zealand waters but caused no further damage or casualties.

To help with the overstretched naval resources, the Women's Royal New Zealand Naval Service (WRNZNS) was formed in April 1942. These women, known as Wrens, worked primarily as signallers, intelligence collectors, book-keepers and drivers. The WRNZNS peaked in 1944 with a strength of 519. In total 640 women served as Wrens during the war.[11]

New Zealand's war at sea also included those serving in the Merchant Navy. The name was archaic, but it reflected the vital importance of these men;

governments of the day regarded their merchant fleets as a fourth military service. The Merchant Navy consisted of all volunteers working on the ships that carried troops, fuel, food and raw materials across the world's oceans. As Neill Atkinson has noted, 'Regardless of their civilian status, merchant seafarers of all nations soon found themselves in the front lines of the war at sea.'[12] Keeping sea lanes open and functioning was vital to the survival of both Britain and New Zealand. No civilians faced greater risks from the enemy than merchant seamen. Of the approximately 3000 New Zealanders who served in the Merchant Navy during the war, at least 120 lost their lives and 123 were captured or interned.[13]

AFTER THE JAPANESE attack on Pearl Harbor on 7 December 1941, which brought the United States into the war, the RNZN operated mainly in the Pacific Ocean. While several thousand New Zealanders continued to serve with the Royal Navy, the RNZN became part of the US Navy's South Pacific Command. Both *Leander* and *Achilles* were used to escort American troops to Guadalcanal in the Solomons and were then used throughout the Solomons Campaign. Both cruisers were badly damaged in this fighting. On 5 January 1943, during a Japanese air raid at Guadalcanal, HMNZS *Achilles* was hit by a bomb. The ship was badly damaged and 13 men were killed. *Achilles* had to return to New Zealand for repairs and did not rejoin South Pacific Command until mid-1944. HMNZS *Leander* suffered worse damage when, on 13 July 1943, the ship was hit by a torpedo, which killed 26 men and left a gaping hole in the cruiser. This damage could not be repaired in New Zealand, so the vessel was sent to Britain and replaced by HMNZS *Gambia*, the largest ship ever to serve in the RNZN.

With the two cruisers out of action for some time, New Zealand's naval efforts were confined to smaller vessels, especially minesweepers and motor launches. Early in 1939 the decision had been made to build a trio of vessels 'as mine-sweeping, gunnery, and torpedo training ships'. Delays meant that these Bird-class corvettes, named HMNZSS *Moa*, *Tui* and *Kiwi*, did not reach New Zealand from Scotland until 1942. A number of civilian vessels in New Zealand were also converted or taken over as minesweepers.[14] Ships of the 25th Minesweeping Flotilla served in the Pacific from the end of 1941 to mid-1945. At the end of October 1942, the flotilla was offered to the US Navy and on 12 December *Matai*, *Kiwi*, *Moa* and *Tui* sailed to the Solomon Islands to take part in the Guadalcanal campaign.[15] In March 1944, 12 launches (motor boats) were sent to the Solomons where they formed the 80th and 81st Motor Launch Flotillas. These were still there at the end

RIGHT | On 27 February 1941, *Leander* encountered the Italian armed merchant raider *Ramb I* in the Indian Ocean and a short, intense naval action followed. *Leander*'s captain, R. H. Bevan, photographed the action from the bridge. Here, *Ramb I* is trying to steam away.

NATIONAL MUSEUM OF THE ROYAL NEW ZEALAND NAVY, OC348

RIGHT | *Ramb I* has been hit and set on fire.

NATIONAL MUSEUM OF THE ROYAL NEW ZEALAND NAVY, OC350

BELOW | *Ramb I* sinking fast after being set ablaze by *Leander*.

NATIONAL MUSEUM OF THE ROYAL NEW ZEALAND NAVY, OC356, OC358

LEFT | One of *Leander*'s 4-inch gun crews.

NATIONAL MUSEUM OF THE ROYAL NEW ZEALAND NAVY, AAG 0071

LEFT | *Leander* stops to pick up the Italian survivors from the sinking of *Ramb I*.

NATIONAL MUSEUM OF THE ROYAL NEW ZEALAND NAVY, AAG 0101

ABOVE | Team sports were one way to combat boredom at sea, a constant concern for commanders. A tug-of-war competition aboard *Leander* somewhere in the Indian Ocean in 1941.

NATIONAL MUSEUM OF THE ROYAL NEW ZEALAND NAVY, 2006.186.148.P28.3

of the war.[16] Later in the year, the British Far Eastern Fleet, which included the New Zealand cruisers *Gambia* and *Achilles* and the corvette *Arbutus*, sailed to the Pacific to take part in what proved to be the final naval operations against Japan. HMNZS *Gambia* was in Tokyo Bay as part of the Allied naval contingent when Japan formally surrendered. The RNZN had certainly earned its place there.

DURING THE WAR, in a reversal of the situation in 1939, the RNZN loaned thousands of men to the Royal Navy. At its peak in July 1945 the RNZN boasted 10,557 personnel, only 5 per cent of whom were on loan from the Royal Navy.[17] Despite this massive growth, the RNZN remained the smallest of the three services, and its casualties were correspondingly relatively light: 740 were killed or wounded and another 460 were killed in action while serving. Many of those who died, though, were very young. You did not have to be 20 years of age to serve with the navy; you could be as young as 15. Some 10 per cent of the RNZN casualties were under 20. The youngest was Boy First Class Henry Honeyfield, who was killed in action on *Achilles* in January 1943. He had just turned 17.[18] Thomas Burke and Edward Walls — both 15 and both from Lower Hutt and serving as deck boys on the merchant ship SS *Port Hunter* — were killed when their munitions-carrying vessel was hit by a German torpedo. The ship sank within minutes, taking 88 men, including seven New Zealanders, with it. Burke and Walls were 'almost certainly the youngest New Zealanders killed in combat during the 20th century'.[19]

Because of its small size, low number of casualties and lack of Victoria Crosses, the RNZN's part in the Second World War is not well known and has undoubtedly been neglected by historians. The RNZN's role was crucial to the war effort. Keeping oceans open to trade was essential to winning the war, which meant defeating the U-boat menace and destroying the surface fleets of Germany, Italy and Japan. The war against Japan, which preoccupied the RNZN from 1942, was primarily a naval one. As Peter Dennerley has suggested, by 1945 the RNZN 'had earned a reputation for getting the job done, no matter what the difficulties or dangers involved'.[20] This should never be forgotten.

ABOVE | HMNZS *Leander* receiving a new coat of paint at Alexandria.

NATIONAL MUSEUM OF THE ROYAL NEW ZEALAND NAVY, AAG 0072

ABOVE | In Alexandria at the end of July 1941 a sailor cleans the barrels of *Leander*'s anti-aircraft guns.

NATIONAL MUSEUM OF THE ROYAL NEW ZEALAND NAVY, OC662

BELOW | 'Tourists' — members of *Leander*'s crew at Nazareth in 1941.

NATIONAL MUSEUM OF THE ROYAL NEW ZEALAND NAVY, 2006.186.148.P40.4

RIGHT | HMS *Hood*, the last battlecruiser built for the Royal Navy, is bracketed with shells fired from the German battleship *Bismarck* on 24 May 1941 during the Battle of the Denmark Strait. Less than a minute later, *Bismarck*'s second salvo hit and sank the *Hood*. Only three of the crew survived.

NATIONAL MUSEUM OF THE ROYAL NEW ZEALAND NAVY, OC1096

LEFT | On 25 November 1941, HMS *Barham* was struck by three torpedoes and sank so quickly, within five minutes, that most of the 862 crew died. This photograph was taken by Lieutenant Stanley Jervis DSC, from Taihape, aboard HMS *Valiant*.

NATIONAL MUSEUM OF THE ROYAL NEW ZEALAND NAVY, OC1296

ABOVE | A British battleship in the Mediterranean opens fire on bombers during a night raid. Despite being clearly illuminated by the muzzle flashes, the ship was not hit.

NATIONAL MUSEUM OF THE ROYAL NEW ZEALAND NAVY, OC933

RIGHT | In January 1944 mine disposal officer Lieutenant E. Savage of the Royal Navy inspects a German mine that has washed up on Cutters Rock in Manukau Harbour. At the time, this mine was thought to have drifted across the Tasman, but hundreds of German mines laid off New Zealand's coast remained undetected during the war.

NATIONAL MUSEUM OF THE ROYAL NEW ZEALAND NAVY, OC143

RIGHT | Inspecting a damaged Royal Naval Air Service bomber in Tanganyika.

NATIONAL MUSEUM OF THE ROYAL NEW ZEALAND NAVY, 2014.32.30.P12.3

BELOW | 'Boy, I earned 'em': Fleet Air Arm pilot 'Pop' Wintringham shows off his hard-earned aviator's wings.

NATIONAL MUSEUM OF THE ROYAL NEW ZEALAND NAVY, 2014.32.23.P56.2

LEFT | Fleet Air Arm pilot Morris Dingwall offers a silent prayer before his first solo flight at Grosse Isle, Canada, in December 1941 while on the Empire Air Training Scheme (EATS).

NATIONAL MUSEUM OF THE ROYAL NEW ZEALAND NAVY, 2014.32.27.P21.2

LEFT | A convoy in sailing order during the Battle of the Atlantic.

NATIONAL MUSEUM OF THE ROYAL NEW ZEALAND NAVY, ACD 0082

ABOVE | The first class of New Zealand pilot trainees for the Fleet Air Arm, photographed in mid-1941.

NATIONAL MUSEUM OF THE ROYAL NEW ZEALAND NAVY, AQD 0122

RIGHT | As the only British harbour between Gibraltar and Alexandria, the island of Malta was an essential staging post for supply convoys to British forces in North Africa. In August 1942, the aircraft carrier *Indomitable* came under air attack while escorting a convoy to Malta. Hit by four bombs and on fire, the vessel was still capable of making 28 knots on the return journey to Gibraltar.

NATIONAL MUSEUM OF THE ROYAL NEW ZEALAND NAVY, OC1100

LEFT | New Zealand Fleet Air Arm pilots in a convoy run to Malta in August 1942. They are named, from left, Hill, Napper, Richardson, Long, Morrison, Reece and F. A. J. Pennington.

NATIONAL MUSEUM OF THE ROYAL NEW ZEALAND NAVY, OC1099

RIGHT | Fleet Air Arm pilots at the Hal Far aerodrome on Malta before setting out on a bomb attack on Gela on 23 August 1942. They are, from left, Sub-Lieutenants C. T. White, E. F. Pratt and Elliott. Colin White DSC was a member of the Royal New Zealand Naval Volunteer Reserve (RNZVR).

NATIONAL MUSEUM OF THE ROYAL NEW ZEALAND NAVY, OC1305

RIGHT | A Fleet Air Arm Swordfish releases its torpedo.

NATIONAL MUSEUM OF THE ROYAL NEW ZEALAND NAVY, AQD 0032

BELOW | A plane coming in to land on HMS *Tracker*.

NATIONAL MUSEUM OF THE ROYAL NEW ZEALAND NAVY, AQD 0079

ABOVE | The New Zealand hospital ship *Maunganui* in Wellington Harbour in March 1945. Capable of carrying 390 patients, the *Maunganui* commenced its sailing duties in April 1941 and made 16 return journeys from Greece, Italy and the Middle East, always with a full complement of patients. Its final voyage as a hospital ship was to take home British prisoners of war in November 1945. During the war the *Maunganui* carried 5677 patients.

NATIONAL MUSEUM OF THE ROYAL NEW ZEALAND NAVY, OC583

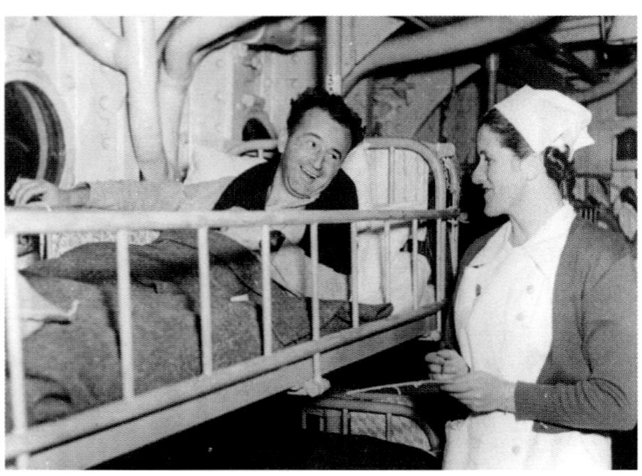

LEFT | A nurse checks on a patient on HMNZHS *Maunganui* in May 1942.

NATIONAL MUSEUM OF THE ROYAL NEW ZEALAND NAVY, OC723

ABOVE | Three minesweepers, led by T155, HMNZS *Inchkeith*, enter Whāngārei Harbour in 1942. The second vessel is T174, *Killegray*.

NATIONAL MUSEUM OF THE ROYAL NEW ZEALAND NAVY, OC244

ABOVE | The remains of the Japanese submarine *I1* sunk off Guadalcanal on 30 January 1943 by the minesweepers *Kiwi* and *Moa*. During the action *Kiwi* rammed the submarine twice.

NATIONAL MUSEUM OF THE ROYAL NEW ZEALAND NAVY, OC344

LEFT | *Kiwi* in dry dock in Auckland with extensive bow damage after the encounter with the Japanese submarine.

NATIONAL MUSEUM OF THE ROYAL NEW ZEALAND NAVY, OC115

ABOVE | *Leander* gun crews resting after the action at Kolombangara in the Solomons in July 1943 during which the ship was hit by a torpedo, which made a massive hole in its side and killed 26 members of the crew.

NATIONAL MUSEUM OF THE ROYAL NEW ZEALAND NAVY, OC806

ABOVE | Some sailors' religious beliefs did not permit burial at sea. Stoker Class 1 Donald Slattery, whose coffin is being carried, and Chief Engineer Room Artificer K. B. Fraser, from the *Leander*, are buried in Auckland's Waikumete Cemetery. An unknown sailor lies beside them.

NATIONAL MUSEUM OF THE ROYAL NEW ZEALAND NAVY, AAG 0316

LEFT | Burials at sea after *Leander*'s action at Kolombangara.

NATIONAL MUSEUM OF THE ROYAL NEW ZEALAND NAVY, OC811

RIGHT | Senior officers inspect *Leander*'s damage while it is in Calliope Dock at HMNZS *Philomel,* Devonport, Auckland. The appalling damage is all too clear.

NATIONAL MUSEUM OF THE ROYAL NEW ZEALAND NAVY, 2013.70.2655

LEFT | A partly repaired HMNZS *Leander* leaving Calliope Dock. Further work could only be completed in Britain.

NATIONAL MUSEUM OF THE ROYAL NEW ZEALAND NAVY, AAG 0032

RIGHT | Signalling an aircraft to land on HMS *Indefatigable*. This photograph was taken in New Zealand waters in December 1945.

NATIONAL MUSEUM OF THE ROYAL NEW ZEALAND NAVY, OC315

BELOW | Fleet Air Arm Fairey Fulmars in mid-flight.

NATIONAL MUSEUM OF THE ROYAL NEW ZEALAND NAVY, 2014.32.31.P4.3

LEFT | Fleet Air Arm pilot Sub-Lieutenant Owen (Lefty) Munro of No. 882 Squadron poses in front of his fighter plane, complete with Kiwi emblem.

NATIONAL MUSEUM OF THE ROYAL NEW ZEALAND NAVY, 2014.32.31.P34.3

ABOVE | A Fleet Air Arm pilot of No. 836 Squadron coming in to land on MV *Adula*, a merchant aircraft carrier.

NATIONAL MUSEUM OF THE ROYAL NEW ZEALAND NAVY, 20144.32.46

LEFT | Five Hellcat fighters of Fleet Air Arm No. 888 Squadron in tight formation.

NATIONAL MUSEUM OF THE ROYAL NEW ZEALAND NAVY, 2014.32.22.P2.1

ABOVE | Two pilots of Fleet Air Arm No. 888 Squadron.

NATIONAL MUSEUM OF THE ROYAL NEW ZEALAND NAVY, 2014.32.22.P26.1

RIGHT | Taking off with low wind speed over the deck. It took nerve and courage to be a Fleet Air Arm pilot.

NATIONAL MUSEUM OF THE ROYAL NEW ZEALAND NAVY, AQD 0094

LEFT | Lieutenant Commander F. Nottingham of Fleet Air Arm, the No. 866 Squadron Commander, briefing air crew.

NATIONAL MUSEUM OF THE ROYAL NEW ZEALAND NAVY, AQD 0069

RIGHT | A Fleet Air Arm advertisement published in the *Southland News*. Many New Zealanders responded to the call.

NATIONAL MUSEUM OF THE ROYAL NEW ZEALAND NAVY, AQD 0053

BELOW | As a Fleet Air Arm plane comes in to land the pilot hopes that the dangling hook will catch one of the trip wires laid across the deck.

NATIONAL MUSEUM OF THE ROYAL NEW ZEALAND NAVY, 2014.32.50

BELOW RIGHT | Caught it — a successful landing is completed.

NATIONAL MUSEUM OF THE ROYAL NEW ZEALAND NAVY, 2014.32.51

JOIN THE FLEET AIR ARM

A GREAT OPPORTUNITY PRESENTS ITSELF TO YOUNG MEN WHOSE AGES RANGE FROM 18 TO 28 YEARS TO JOIN THE NAVY AND LEARN TO FLY!

Pilots and Observers (Navigators) are urgently required for service overseas for the duration of hostilities. Candidates should be single, or, if married, with not more than one child.

A secondary education and physical fitness for flying duties including perfect eyesight and colour vision are essential.

Applications will be accepted from such men to enter the Royal New Zealand Navy as Naval Airmen with a view to advancement to Petty Officer or promotion to Commissioned rank after training in the United Kingdom.

For further particulars, conditions of service pay, etc., write to the nearest Naval recruiter:—

Naval Base, Devonport, AUCKLAND; 164 Featherston Street, WELLINGTON; R.N.Z.N.V.R. Headquarters, St. Asaph Street, CHRISTCHURCH; or 6 Castle Street DUNEDIN.

ABOVE | No. 888 Squadron Fleet Air Arm pilot Alan Sharpe from Whanganui has missed his landing. He survived this crash, but was later killed in action over France on 19 August 1944.

NATIONAL MUSEUM OF THE ROYAL NEW ZEALAND NAVY, 2014.32.31.P43.2

ABOVE | The moment of impact as a Hellcat fighter hits the deck of an aircraft carrier after crashing into its barrier. The pilot can be seen holding himself away from the cockpit's dashboard to prevent head injuries. The aircraft was poised like this for a second or two before falling backwards.

NATIONAL MUSEUM OF THE ROYAL NEW ZEALAND NAVY, OC962

BELOW | Some New Zealanders who served in the Royal Navy went on to have distinguished careers in the military, in politics and in the civil service. Captain Peter Phipps, here at the compass on the bridge of HMNZS *Royalist*, rose to the highest command in the navy and the defence force.

NATIONAL MUSEUM OF THE ROYAL NEW ZEALAND NAVY, 2013.70.3407

RIGHT | HMNZS *Kiwi* under way in the Hauraki Gulf.

NATIONAL MUSEUM OF THE ROYAL NEW ZEALAND NAVY, 2013.70.1169

BELOW | A rating keeps a sharp lookout as HMNZS *Killegray* sweeps for mines between Auckland and Whāngārei.

NATIONAL MUSEUM OF THE ROYAL NEW ZEALAND NAVY, OC365

ABOVE | The minesweeper *Killegray* trains for its role.

NATIONAL MUSEUM OF THE ROYAL NEW ZEALAND NAVY, 2006.186.157.2

RIGHT | One way of clearing a minefield — a controlled explosion in 1943.

NATIONAL MUSEUM OF THE ROYAL NEW ZEALAND NAVY, 2006.186.157.1

LEFT | Manning the anti-aircraft gun of harbour defence motor launch *Q1189* in 1944.

NATIONAL MUSEUM OF THE ROYAL NEW ZEALAND NAVY, ABF 0022

ABOVE | New Zealand High Commissioner Bill Jordan pays a visit to the crew of *Gambia* at Portsmouth in January 1944.

NATIONAL MUSEUM OF THE ROYAL NEW ZEALAND NAVY, OC45

ABOVE | Fleet Air Arm pilots receive a briefing before their attack on the German battleship *Tirpitz* in April 1944.

NATIONAL MUSEUM OF THE ROYAL NEW ZEALAND NAVY, 1291

RIGHT | Five New Zealand Fleet Air Arm pilots from HMS *Emperor* who took part in the *Tirpitz* attack are, from left, SubLieutenants Scanes from Christchurch, Cranwell from Auckland, McLennan from Thames and Jellie from Auckland, and Lieutenant Hill from Wellington.

NATIONAL MUSEUM OF THE ROYAL NEW ZEALAND NAVY, OC340

ABOVE | Trucks and infantry loaded on a landing craft ready to move onto the beach in Normandy in June 1944 as part of Operation 'Overlord'.

NATIONAL MUSEUM OF THE ROYAL NEW ZEALAND NAVY, OC1049

ABOVE | Glider-borne troops cross the English Channel above ships of the Royal Navy on the eve of Operation 'Overlord', the landing on the Normandy beaches. Almost 5000 New Zealanders were serving in the Royal Navy in mid-1944.

NATIONAL MUSEUM OF THE ROYAL NEW ZEALAND NAVY, OC857

LEFT | Fleet Air Arm pilots prepare to provide air cover for soldiers landing on the Normandy beaches on 6 June 1944.

NATIONAL MUSEUM OF THE ROYAL NEW ZEALAND NAVY, 2014.32.49

ABOVE | Lieutenant William Dimes (RNZNVR) from Wellington climbs down from his Wildcat fighter after helping to destroy a Junkers 290 long-range bomber that had been shadowing the convoy protected by the escort carrier HMS *Biter*.

NATIONAL MUSEUM OF THE ROYAL NEW ZEALAND NAVY, OC1294

ABOVE | Just one of thousands of New Zealanders who served in the Royal Navy during the war, Lieutenant Howard Mallitte (RNZNVR) from Wellington watches as his ship sweeps for mines in the Mediterranean in August 1944.

NATIONAL MUSEUM OF THE ROYAL NEW ZEALAND NAVY, OC689

RIGHT | Part of a fleet of seven British aircraft carriers passing through the Straits of Messina in late 1944.

NATIONAL MUSEUM OF THE ROYAL NEW ZEALAND NAVY, 2014.32.26.P34.1

RIGHT | Deck crew moving a Supermarine Seafire, the naval version of the Spitfire, on board HMS *Hunter* in August 1944. The Seafire was part of No. 807 Squadron and was used in support of Operation 'Dragoon', the invasion of Southern France in August 1944.

NATIONAL MUSEUM OF THE ROYAL NEW ZEALAND NAVY, 2014.32.26.P28.1

ABOVE | Inspecting the ammunition belts of the planes of Fleet Air Arm No. 807 Squadron, which was supporting Operation 'Dragoon'.

NATIONAL MUSEUM OF THE ROYAL NEW ZEALAND NAVY, 2014.32.26.P29.1

BELOW | A montage of the pilots of Fleet Air Arm Squadron 807 in August 1944.

NATIONAL MUSEUM OF THE ROYAL NEW ZEALAND NAVY, 2014.32.26.P14.1

ABOVE | A member of Fleet Air Arm No. 888 Squadron examines aerial photographs.

NATIONAL MUSEUM OF THE ROYAL NEW ZEALAND NAVY, 2014.32.22.P38.1

ABOVE | A Fleet Air Arm Wildcat fighter in mid-flight.

NATIONAL MUSEUM OF THE ROYAL NEW ZEALAND NAVY, 2014.32.24.P32.1

LEFT | Grumman Hellcat fighters of Fleet Air Arm No. 888 Squadron in tight formation.

NATIONAL MUSEUM OF THE ROYAL NEW ZEALAND NAVY, 2014.32.22.P36.1

LEFT | An aircraft carrier with its planes in the STO-wing position in the foreground. The STO-wing was a wing-folding mechanism developed early in 1941.

NATIONAL MUSEUM OF THE ROYAL NEW ZEALAND NAVY, 2014.32.31.P44.4

RIGHT | HMS *Wren* picks up the pilot and air gunner from a crashed plane. The observer was not recovered.

NATIONAL MUSEUM OF THE ROYAL NEW ZEALAND NAVY, AQD 0086

LEFT | A Supermarine Seafire of the Fleet Air Arm No. 888 Squadron crash lands on HMS *Battler*.

NATIONAL MUSEUM OF THE ROYAL NEW ZEALAND NAVY, 2014.32.22.P29.1

RIGHT | New Zealanders serving on board the escort carrier *Empress* in June 1945 pose with the ship's mascot. Back row from left: A. R. Sutton, C. I. McEwan, R. Braithwaite and F. Rogers. Middle row: K. R. Stevens, G. Cronin and J. W. Jenkins. A. N. Burr is in front with the ship's cat.

NATIONAL MUSEUM OF THE ROYAL NEW ZEALAND NAVY, OC 328

BELOW | Two German naval ratings standing on the damaged port propeller shaft of the capsized *Tirpitz* in a Norwegian fjord. The battleship had been sunk by Lancaster bombers of the RAF on 13 November 1944. This photograph was taken by a low-flying reconnaissance plane in early 1945.

NATIONAL MUSEUM OF THE ROYAL NEW ZEALAND NAVY, OC341

ABOVE | This landing did not go to plan either.

NATIONAL MUSEUM OF THE ROYAL NEW ZEALAND NAVY, 2014.32.40

LEFT | The wash between HMNZS *Gambia* and a tanker while refuelling at sea would be rated 'moderate to rough'; it required excellent seamanship to complete this operation.

NATIONAL MUSEUM OF THE ROYAL NEW ZEALAND NAVY, AAI 0146

RIGHT | A gun crew of HMNZS *Gambia* at action stations during her work-up in the Hauraki Gulf in January 1945. The crew are carrying 4-inch artillery rounds.

NATIONAL MUSEUM OF THE ROYAL NEW ZEALAND NAVY, OC 298

LEFT | On 4 May 1945, the aircraft carrier HMS *Formidable* was extensively damaged when hit by a Japanese kamikaze suicide plane in the action off the Sakishima Islands.

NATIONAL MUSEUM OF THE ROYAL NEW ZEALAND NAVY, OC267

ABOVE | Transferring stores between *Achilles* and *Arbutus* in August 1945, 300 miles from the coast of Japan, was quite an occasion. Both ships' companies line their respective decks to exchange greetings and catch up on news.

NATIONAL MUSEUM OF THE ROYAL NEW ZEALAND NAVY, OC845

RIGHT | HMNZS *Gambia* firing a 6-inch broadside at Kamaishi on 9 August 1945. This was the last naval bombardment of the war.

NATIONAL MUSEUM OF THE ROYAL NEW ZEALAND NAVY, AAI 0112

LEFT | VJ Day celebrations on 15 August 1945 seen from the deck of a British aircraft carrier.

NATIONAL MUSEUM OF THE ROYAL NEW ZEALAND NAVY, 2014.32.54

LEFT | A New Zealand landing party at Yokosuka, Japan, in 1945.

NATIONAL MUSEUM OF THE ROYAL NEW ZEALAND NAVY, AAI 0163

ABOVE | Two naval ratings from HMNZS *Gambia* stand guard while an American officer examines a range-finder at the entrance to Tokyo Bay.

NATIONAL MUSEUM OF THE ROYAL NEW ZEALAND NAVY, OC815

ABOVE | Peace at last. The Allied occupation fleet in Sagami Bay on 27 August 1945. Mount Fuji is clearly visible on the shoreline.

NATIONAL MUSEUM OF THE ROYAL NEW ZEALAND NAVY, AAI 0101

ABOVE | A fire crew spray foam on a burning New Zealand Corsair fighter. This photograph was taken on Green Island in March 1945.

AIR FORCE MUSEUM OF NEW ZEALAND, PR5230

| CHAPTER 7 |

The Air War over the Pacific

As well as sending several thousand men to fight with the RAF, the RNZAF trained many more in New Zealand who were then deployed to the Pacific, where they fought alongside their American and Australian allies.

From late 1941 through to the end of 1942, as the Japanese advanced steadily through the Pacific, the home defence of New Zealand was also a vital consideration. In 1941, in response to a request from New Zealand for some modern aircraft to respond to the looming Japanese threat, Britain supplied 36 Hudson bombers. That year, too, New Zealand created the Women's Auxiliary Air Force (WAAF) to widen the pool of talent it could use. Almost 5000 WAAFs, as they were known, served during the war years.[1] When the Japanese bombed Pearl Harbor, the RNZAF had just 641 aircraft in New Zealand, most of them trainers. Only the recently acquired Hudson bombers were up-to-date and combat-worthy. But all available aircraft in New Zealand were pressed into service under the Forces Available for Anti-Invasion scheme (FAFAI). These included such antiquated aircraft as Tiger Moths, Vildebeests, Baffins and Vincents. From then on there was a huge build-up of air assets and an increase in personnel.

It was the RNZAF that made New Zealand's main contribution to the Pacific theatre, where it provided air support for several significant land campaigns. No. 488 Squadron, 155 strong and the first fighter squadron formed in New Zealand, took part in the disastrous Singapore and Malaya campaign, flying old and poorly performing Brewster Buffalo fighters. The squadron had only just arrived in Singapore and was still training its pilots when the Japanese attacked Pearl Harbor. When Singapore fell on 15 February 1942 most of the RNZAF personnel were evacuated, but some were taken prisoner and endured long years of brutal captivity. An Aerodrome Construction Squadron of RNZAF also served in this campaign, building airfields in southern Malaya. Around 400 strong, and consisting primarily of men recruited from the Public Works Department and private construction companies, the unit had been the first RNZAF presence in the region, arriving there in October 1941.[2]

After this catastrophe New Zealand appealed urgently to Britain and the United States for newer and better fighter planes to be made available to the RNZAF.

As a result the first Kittyhawk fighters arrived in New Zealand in March 1942. No. 14, the first New Zealand-based squadron, flying the Kittyhawks, was formed at Masterton, and more soon followed. The RNZAF fighter squadrons, equipped with Kittyhawks, and bomber squadrons with the Hudson bombers, were first used during the Solomons campaign in 1942, to support American ground operations. In July that year, based in New Caledonia, No. 9 Squadron began flying anti-submarine patrols with its Hudson bombers. Three months later another RNZAF bomber squadron, No. 3, joined it on anti-shipping and anti-submarine patrol operations. No. 3 Squadron was initially based at Espiritu Santo, but from November 1942 moved to Henderson Field on Guadalcanal. In early 1943, other RNZAF units soon joined these two squadrons, including a repair unit on Espiritu Santo, a group headquarters and newly formed fighter, bomber and transport squadrons. On 2 April 1943, a No. 3 Squadron Hudson shot down a Japanese floatplane. This was the first RNZAF air combat victory of the Solomons campaign.[3]

As THE FIGHTING progressed, new and better planes were made available to the RNZAF. As well as the Kittyhawks, which struggled to match the Japanese Zero, and Hudson bombers, in 1944 RNZAF pilots were flying a range of new planes including Venturas, Corsairs, Avengers, Dauntless dive-bombers and a variety of transport aircraft. At the end of 1943 the RNZAF had enough squadrons and pilots operating in the Solomons to establish the New Zealand Fighter Wing at New Georgia. It ended the war with a record of 99 destroyed aircraft and another 24 probably destroyed. When the war ended in August 1945 the RNZAF had over 7000 personnel on operations in the South West Pacific area.[4]

Sadly, New Zealand's impressive air strength did not endure long after the war. After reaching its highest personnel numbers in 1944, by 1946 the strength of the RNZAF was just 5300.[5] Most of the wartime operational squadrons were disbanded by the end of 1945 and hundreds of aircraft were sold off or condemned to the scrapheap. But from late 1941 through to VJ Day in August 1945, the RNZAF in the Pacific 'played a small but important, and at times, vital part in the campaigns'.[6] Its contribution to the Allied war effort and to the defeat of Japan cannot be denied.

RIGHT | At Tebrau in Malaya No. 1 Aerodrome Construction Squadron men prepare a mine that will destroy much of the airfield they have been working on. Although this was a frustrating task, it was preferable to allowing the Japanese to profit from their hard work.

AIR FORCE MUSEUM OF NEW ZEALAND, ALB161705110

LEFT | Members of No. 1 Aerodrome Construction Squadron waiting for transport to evacuate them from Malaya to Singapore.

AIR FORCE MUSEUM OF NEW ZEALAND, ALB161705109

RIGHT | Lowering a mine that would leave a huge crater, making the airfield unusable.

AIR FORCE MUSEUM OF NEW ZEALAND, ALB161705111

LEFT AND ABOVE | A burning wreckage and a bullet-riddled car after a Japanese air raid at RAF Station Kallang, Singapore, 1941.

AIR FORCE MUSEUM OF NEW ZEALAND, ALB921345118, ALB92145115

RIGHT | Personnel working on a No. 488 Squadron Hurricane. The location is thought to be RAF Station Kallang, Singapore, in 1941.

AIR FORCE MUSEUM OF NEW ZEALAND, ALB921345119

ABOVE | The Singapore docks burn after a Japanese air raid on 2 February 1942.

AIR FORCE MUSEUM OF NEW ZEALAND, ALB161705132

BELOW | Waiting to board SS *Darvel* to escape from Singapore.

AIR FORCE MUSEUM OF NEW ZEALAND, ALB161705145

ABOVE | The wreckage of a No. 243 Squadron Brewster Buffalo fighter flown by Sergeant Pilot Reginald Newman from Wellington. On 12 January 1942, Newman flew to intercept a formation of Japanese bombers over Malaya, but before he reached them was attacked by 15 Japanese fighter planes. He had no chance of surviving this encounter.

AIR FORCE MUSEUM OF NEW ZEALAND, ALB161705102

ABOVE | Members of No. 1 Aerodrome Construction Squadron on board SS *Darvel* after being evacuated from Singapore. Shortly after this photograph was taken the ship was attacked by Japanese bombers and several men were hit by bomb fragments. The only fatality was the man on the left, Leading Aircraftman Albert Stanford from Huiroa in Taranaki, who was buried at sea on 11 February 1942.

AIR FORCE MUSEUM OF NEW ZEALAND, ALB161705104

ABOVE | Leading Aircraftman Colin Kayser from Raetihi, of No. 1 Aerodrome Construction Squadron, described on the back of the photograph as 'the most popular man of the unit'. He died of wounds sustained during a Japanese bomb attack on Singapore on 3 February 1942.

AIR FORCE MUSEUM OF NEW ZEALAND, ALB161705105

LEFT | No. 14 Squadron pilots race to their Kittyhawks during a training exercise in New Zealand.

AIR FORCE MUSEUM OF NEW ZEALAND, PR754

ABOVE | Six No. 16 Squadron Curtiss P-40 Kittyhawks and a No. 9 Squadron Lockheed Hudson bomber on the way to Guadalcanal in 1943.

AIR FORCE MUSEUM OF NEW ZEALAND, PR1172

ABOVE | Pilots practise rapid evacuation from a Carley float.

KEN LUMSDEN, PAUL LUMSDEN COLLECTION

RIGHT | Members of a maintenance unit pose with Allison V-12 aircraft engines, probably at Guadalcanal.

AIR FORCE MUSEUM OF NEW ZEALAND, ALB921345265

ABOVE | A No. 67 Squadron Hurricane flown by William Christiansen had endured a belly landing at RAF Station Alipore, India, in April 1942. Christiansen was killed in action in 1943.

AIR FORCE MUSEUM OF NEW ZEALAND, ALB902763A185

ABOVE | RNZAF armourers about to arm a Kittyhawk fighter.

KEN LUMSDEN, PAUL LUMSDEN COLLECTION

LEFT | A P-40 Kittyhawk taking off for a mission. This is most likely to be a No. 14 Squadron aircraft operating at Kukum Field, Guadalcanal.

AIR FORCE MUSEUM OF NEW ZEALAND, PR1228

RIGHT | Ground crew servicing P-40 Kittyhawks at Kukum Field, Guadalcanal, in 1943.

AIR FORCE MUSEUM OF NEW ZEALAND, PR1252

LEFT | Maintenance Unit men working on the engine of a RNZAF plane, either a Vildebeest or a Vincent.

AIR FORCE MUSEUM OF NEW ZEALAND, ALB87017085

ABOVE | The RNZAF ensign is hoisted by Sergeant Tommy Grant, most likely at Guadalcanal in July 1943.

AIR FORCE MUSEUM OF NEW ZEALAND, PR1170

LEFT | A No. 14 Squadron Kittyhawk breaks formation during a training flight, probably near RNZAF Station Whenuapai in 1943.

AIR FORCE MUSEUM OF NEW ZEALAND, PR746

BELOW | Ground crew servicing a No. 14 Squadron Kittyhawk before the squadron's departure from Whenuapai in September 1943.

AIR FORCE MUSEUM OF NEW ZEALAND, PR901

BELOW | A member of No. 14 Squadron checks his Luger pistol before leaving Whenuapai in 1943.

AIR FORCE MUSEUM OF NEW ZEALAND, PR887

ABOVE | Two Hudsons of No. 3 Squadron on patrol over Guadalcanal in July 1943.

AIR FORCE MUSEUM OF NEW ZEALAND, 1987-375-18

ABOVE | Ground crew, most likely from No. 1 Fighter Maintenance Unit at Kukum Field, Guadalcanal, in 1943, servicing an Allison V-12 aero engine. A Kittyhawk fighter is in the background.

AIR FORCE MUSEUM OF NEW ZEALAND, PR1258

RIGHT | Loading 300-pound depth charges on board the Hudson bombers of No. 9 Squadron, probably at Pallikub Field on Espiritu Santo in 1943.

AIR FORCE MUSEUM OF NEW ZEALAND, PR1384

ABOVE | A Corsair practises skip bombing at Cape Esperance, Guadalcanal, in 1943. The beached Japanese ship, the *Yamazuki Maru*, was much photographed by Allied troops.

AIR FORCE MUSEUM OF NEW ZEALAND, ALB 86260060

RIGHT | Ready to fly his Kittyhawk in 1943, No. 14 Squadron pilot P. S. Green gives the thumbs up signal.

AIR FORCE MUSEUM OF NEW ZEALAND, PR761

LEFT | No. 14 Squadron pilot Flying Officer Geoffrey Fisken beside his P-40 Kittyhawk, the 'Wairarapa Wild Cat', at Guadalcanal in July 1943. As the Japanese flags indicate, he had 11 confirmed 'kills'.

AIR FORCE MUSEUM OF NEW ZEALAND, PR1162

ABOVE | In November 1943, No. 18 Squadron officers pose in front of the New Zealand Fighter Wing scoreboard at Ondonga airfield, New Georgia, which shows 57 kills. From left are R. H. Balfour, J. A. Oldfield, C. D. A. Highet, B. W. Williams and R. W. Bremner.

AIR FORCE MUSEUM OF NEW ZEALAND, PR2367

RIGHT | Armourers load ammunition into the port wing magazines of a P-40 Kittyhawk that was most likely part of No. 18 Squadron at Torokina, Bougainville, in the Solomons, in 1944.

AIR FORCE MUSEUM OF NEW ZEALAND, PR2685

RIGHT | A ship burns off the coast of a Pacific island. By early 1944 the Allies controlled the region's skies.

AIR FORCE MUSEUM OF NEW ZEALAND, ALB843302047

ABOVE | Ground crew perform a 320-hour maintenance of a Corsair fighter at Mokerang, Papua New Guinea, in 1944.

AIR FORCE MUSEUM OF NEW ZEALAND, PR4875

RIGHT | No. 6 Squadron personnel with a Catalina moored in the shallows of Havalo Bay, Solomon Islands, in 1944.

AIR FORCE MUSEUM OF NEW ZEALAND, PR3321

ABOVE | A beached Japanese cargo vessel, probably at Guadalcanal in early 1944.

AIR FORCE MUSEUM OF NEW ZEALAND, PR4251

RIGHT | No. 31 Servicing Unit examines Corsair fighters at Espiritu Santo in 1944.

AIR FORCE MUSEUM OF NEW ZEALAND, PR4372

ABOVE | The RNZAF presence at Guadalcanal in 1944 was certainly noticeable.

AIR FORCE MUSEUM OF NEW ZEALAND, ALB87017047

BELOW | 'To live in the hearts we leave behind is not to die', reads the annotation to this photograph. The temporary grave is of No. 18 Squadron pilot flight Sergeant Frederick Donaldson at Guadalcanal. He was later buried in the Bourail New Zealand war cemetery in New Caledonia.

AIR FORCE MUSEUM OF NEW ZEALAND, ALB11212315044

BELOW | An RNZAF guard of honour standing to attention. This is believed to be for the funeral of 23-year-old No. 18 Squadron pilot Flight Sergeant Frederick Arthur Donaldson, who was killed at Guadalcanal on 27 May 1944.

AIR FORCE MUSEUM OF NEW ZEALAND, ALB11212315043

RIGHT | At Torokina, Bougainville, in early 1944, No. 18 Squadron pilots await transport to their aircraft.

AIR FORCE MUSEUM OF NEW ZEALAND, PR2680

ABOVE | Two RNZAF personnel at RNZAF Camp Hinemoa, Bougainville, in 1944.

AIR FORCE MUSEUM OF NEW ZEALAND, ALB87017087

RIGHT | The Electrical Section of a New Zealand maintenance unit at Bougainville in 1944.

AIR FORCE MUSEUM OF NEW ZEALAND, ALB87017095

ABOVE | A line of vehicles and drivers of the Transport Section in Fiji.

AIR FORCE MUSEUM OF NEW ZEALAND, ALB030131003

LEFT | WAAF driver Edith Josephine (Josie) McGovern with the commanding officer's Ford Car at Suva Point, Fiji, in November 1944.

AIR FORCE MUSEUM OF NEW ZEALAND, MUS110157

RIGHT | A hurricane in Fiji on 17 February 1941 killed four people and injured another 80. The frames of the flimsy shelters at Nausori airfield survived the ordeal.

AIR FORCE MUSEUM OF NEW ZEALAND, ALB875584036

RIGHT | A group of RNZAF personnel display captured Japanese equipment at Bougainville in 1944. Wing Commander C. W. K. Nicholls holds the rifle, Flight Lieutenant P. T. Curran the left edge of the flag.

AIR FORCE MUSEUM OF NEW ZEALAND, PR3403

ABOVE | Pilots attend a briefing on Bougainville in 1945.

AIR FORCE MUSEUM OF NEW ZEALAND, PR5740

RIGHT | A record bomb load for No. 2 Squadron's Lockheed Venturas at Green Island in March 1945.

AIR FORCE MUSEUM OF NEW ZEALAND, PR5275

ABOVE | A real team effort — Nos 1 and 25 Servicing Units arrive at Augusta Bay, Bougainville, in April 1945.

AIR FORCE MUSEUM OF NEW ZEALAND, PR5731

RIGHT | After a Corsair pilot at Los Negros crashed when catching a side wind on take-off, firemen douse the plane and pilot with foam. This is likely to have occurred on Bougainville in 1945.

AIR FORCE MUSEUM OF NEW ZEALAND, ALB023161040

ABOVE | A burnt-out Corsair after an accident on 13 April 1945 at Piva airfield on Bougainville.

AIR FORCE MUSEUM OF NEW ZEALAND, PR5739

RIGHT | The Japanese surrender at Bougainville on 20 August 1945.

AIR FORCE MUSEUM OF NEW ZEALAND, PR7260

ABOVE | New Zealand soldiers arriving in theatre. Note the anti-aircraft guns.

ENGINEER CORPS MEMORIAL CENTRE, BISHOP 041

| CHAPTER 8 |

The Ground War in the Pacific

When the long-anticipated Pacific War erupted in December 1941, the bulk of New Zealand's land forces were in the Middle East, where they had just had their first, somewhat Pyrrhic, success of the war in Operation 'Crusader'. In the Pacific theatre New Zealand had only a brigade group and two reconnaissance squadrons in Fiji, and an infantry company on Fanning Island. The New Zealand government, committed to focusing their efforts in the region, soon reinforced 2 NZEF in the Pacific to a two-brigade infantry division that later became 3 New Zealand Division.

The government's preferred option was to have the whole of the NZEF, including 2 New Zealand Division, transferred to the Pacific once Egypt was no longer threatened by Axis forces. As early as March 1942 Prime Minister Peter Fraser warned General Freyberg that New Zealand efforts in the Middle East would have to be reduced.[1] This resulted in 2 NZEF and 2 New Zealand Division not receiving reinforcements at all in 1942, these men instead being used to build up 3 New Zealand Division. Over the next two years, New Zealand three times raised the question of returning its land forces to fight in the Pacific. Each time it was dissuaded with compelling arguments by both Britain and the United States. Only Australia pressed hard for the return of New Zealand's land forces, as it had done earlier with the Australian Imperial Force. The Australian government, and especially its prime minister, John Curtin, was extremely annoyed when New Zealand accepted the views of its key allies and kept its main land component fighting in Italy.[2] This meant that 3 New Zealand Division was never New Zealand's principal land effort. It was always understrength and could not fight large set-piece or lengthy engagements that would consume its precious manpower.

In early 1942, the Americans took responsibility for the defence of Fiji, Tonga and Fanning Island, which allowed 3 Division to return to New Zealand to train for amphibious operations in the Pacific. When the Americans began their long, hard operations to capture the Solomon Islands from Japan, beginning with their assault on Guadalcanal in August 1942, New Zealand soldiers moved back into this theatre of war. The Solomons were the outer bastion of Japan's southward expansion and heavily defended. Initially New Zealand supplied garrison troops for New Caledonia,

Norfolk Island and Tonga, but by the end of the year 3 New Zealand Division was fully established at New Caledonia and ready to participate in operations. It was not until August–September 1943 that 3 Division, consisting of 8 and 14 Brigades and a divisional headquarters, moved to the active war zone of Guadalcanal in the Solomons.

On 18 September 1943, 14 Brigade and divisional headquarters were sent to Vella Lavella in the north-west Solomons to relieve the US Marines who had landed there the previous month. It was the New Zealanders' task to continue fighting the Japanese and clear the island. It took until 9 October before Vella Lavella was secured. In this initial fighting, despite having only a 'limited task' to perform, New Zealand lost 32 men. It was a tough introduction to jungle warfare, but 14 Brigade had performed 'very efficiently' in an 'unfamiliar framework'.[3]

On 27 October, 8 Brigade carried out amphibious landings at the Treasury Islands of Mono and Stirling as part of a mixed New Zealand–American force. This was a preliminary operation before a much larger landing at Bougainville. The two Treasury Islands were needed to establish a radar station that could cover the airspace and sea around Bougainville. When the New Zealanders went ashore at Mono Island on 27 October they were taking part in the first amphibious landing by New Zealand soldiers since Gallipoli in April 1915.[4] The Japanese defenders, unable to prevent the New Zealand and American soldiers landing, were steadily pushed back. Once ashore, the New Zealanders and the Americans quickly established the radar station while engineers built an airfield on Stirling Island. By 12 November both islands were secured. In the fortnight of jungle fighting 40 New Zealanders had been killed and 145 wounded.[5]

The division's last operation started on 15 February 1944 when 14 Brigade, divisional headquarters and attached American units captured Nissan, part of the Green Islands north of Bougainville.[6] Although still only a brigade-sized operation, this was the largest of the three amphibious landings undertaken by New Zealanders in the Pacific: more than 6000 New Zealanders and Americans were put ashore on the first day.[7] Although Nissan Island was the biggest of the Green Islands, it was held only lightly by the Japanese and easily captured. The fighting was over within three days and casualties were light. Ten New Zealanders were killed and 21 wounded.[8]

SERVING IN THE tropical Pacific may not have sounded too strenuous to soldiers slugging their way up the Italian peninsula, and places like Singapore, the Solomon Islands and even Japan 'had an exotic quality for New Zealanders in the early 1940s'.[9] The reality was very different. Added to the dangers of a skilled, brutal foe fanatical in the defence of their country's empire were many other hazards, including all the risks associated with

ABOVE | A New Zealand artillery Quad passes the saluting base at the Joan of Arc Ceremonial Parade at Necal, Noumea. The American theatre commander Admiral William 'Bull' Halsey is third from the left in the front row.

ENGINEER CORPS MEMORIAL CENTRE, ECMC 10952

LEFT | The sinews of war — unloading fuel from a landing craft.

ENGINEER CORPS MEMORIAL CENTRE, BISHOP 039

ABOVE | Soldiers of 3 New Zealand Division at Guadalcanal in August 1943. After months of training, the New Zealand soldiers in the Pacific were in an active war zone.

NATIONAL ARMY MUSEUM, 1993-1317

RIGHT | A New Zealand anti-aircraft gun crew in position with ammunition loaded and ready for action.

NATIONAL ARMY MUSEUM, 1994-3037

LEFT | A New Zealand anti-aircraft gun crew watches the skies in the Solomons.

NATIONAL ARMY MUSEUM, 1994-3037

a hot, damp tropical environment. There was disease such as malaria, dengue fever and elephantiasis, plus spiders, snakes, scorpions, leeches, tropical worms and nipping land crabs. Fighting in the Pacific was definitely no soft option.

By 1944 New Zealand simply did not have enough men to commit to the navy, air force and home defence as well as maintain two infantry divisions on operations. Something had to give. In early January the government decided that one of the infantry divisions would have to be withdrawn and used to reinforce the other. It sought advice from its key allies as to which it should be.[10] In April the decision was made to wind down 3 Division once the Solomons campaign was completed. Initially a cadre staff was left at New Caledonia that could rebuild the division if it were needed. It was not, and on 11 September the War Cabinet decided to disband 3 Division completely.[11] Its headquarters in New Zealand closed on 20 October 1944. New Zealand's role in the Pacific would be left to its numerous air force and navy units operating there.

NEW ZEALAND'S LAND EFFORTS in the Pacific, though small, are of historical significance. This was the first time that New Zealand soldiers fought without the presence of soldiers or commanders from Britain. This presented New Zealand and the 3 New Zealand Division commander Major-General Harold Barrowclough with 'a dilemma'. His division 'was an isolated British formation in an American theatre of war which was primarily naval in operation'.[12] At one point, Barrowclough found himself in command of more than 10,000 American soldiers as well as his 6000 New Zealanders, in what has been described as 'a model of New Zealand–American co-operation in the field'. The Solomons campaign was also the only one in which all three armed forces of New Zealand participated as designated units from their country of origin.[13]

The war in the Pacific has always taken second place to the events that occurred on the other side of the world. The British and American decision taken before Pearl Harbor, and affirmed shortly after the event, to 'beat Germany first' certainly contributed to this perception. As one historian astutely noted of New Zealand's efforts here, 'it may well be that posterity will see a greater significance than we do now in this, the defeat of the first attempt by a major maritime power to extend its sphere of domination southward'.[14] Oliver Gillespie, who wrote the official history of New Zealand's Pacific war, believed that although the taking of Guadalcanal would 'be regarded by future historians as one of the critical battles of history', he too felt that 'the importance of the Pacific war has been insufficiently understood . . . Not until the years adjust them to their proper perspective will the Pacific campaign be fitted correctly into its place in the global war of 1939–45.'[15] This has still not happened.

ABOVE | This photograph of three US Marines, one operating a flame-thrower against a Japanese bunker, indicates the intensity of the fighting in the Pacific and how close the enemy could be. This was taken in Bougainville in either late 1943 or early 1944. The Japanese held on there until the end of the war.

AIR FORCE MUSEUM OF NEW ZEALAND, ALB873251064

BELOW | This Japanese field gun position has been overrun.

AIR FORCE MUSEUM OF NEW ZEALAND, ALB873251071

ABOVE | New Zealand's forward-base post office on Guadalcanal in August–September 1943.

NATIONAL ARMY MUSEUM, 2000-586-1

ABOVE | A makeshift camp in the jungle.

ENGINEER CORPS MEMORIAL CENTRE, BISHOP 027

ABOVE | New Zealand soldiers about to deliver the mail via an air drop.

NATIONAL ARMY MUSEUM, 2000-586-1

ABOVE | Post office staff sorting the mail on Guadalcanal.

NATIONAL ARMY MUSEUM, 2000-586-1

LEFT | General Freyberg inspects troops in the Pacific on his journey to New Zealand in 1943.

ENGINEER CORPS MEMORIAL CENTRE, IMG 0063

RIGHT | The New Zealanders establish a jungle camp in the Solomons. Note the camouflage netting in the bottom left of the image.

NATIONAL ARMY MUSEUM, 2000-586-2

ABOVE | New Zealand sappers preparing timber for construction work.

ENGINEER CORPS MEMORIAL CENTRE, BISHOP 030

BELOW | A New Zealand signaller erecting a wireless mast on Espiritu Santo. Did he use the opportunity to grab a few of the coconuts on offer?

NATIONAL ARMY MUSEUM, 2000-586-1

ABOVE | A soldier takes the opportunity to darn his socks. The tropics were hard on soldiers and their equipment.

NATIONAL ARMY MUSEUM, 1994-3692

ABOVE | Sick New Zealand soldiers await medical evacuation. The men were fighting in a region where tropical diseases were a constant danger.

NATIONAL ARMY MUSEUM, 1994-3037

RIGHT | The evacuation scene is a busy one, with those on stretchers about to be lifted aboard waiting barges and pyjama-clad wounded men being shown the way.

ENGINEER CORPS MEMORIAL CENTRE, BISHOP 036

RIGHT | Soldiers advance cautiously behind a vehicle on a jungle track.

NATIONAL ARMY MUSEUM, 1994-3692

BELOW | A soldier's lonely grave in the jungle.

NATIONAL ARMY MUSEUM, 1994-3692

RIGHT | American and New Zealand soldiers interrogate Japanese prisoners of war. Few Japanese were captured during the Solomons campaign.

NATIONAL ARMY MUSEUM, 1994-3037

RIGHT | New Zealand soldiers pose with their Japanese war trophies. Swords were particularly highly prized.

NATIONAL ARMY MUSEUM, 1994-3037

BELOW | United States soldiers and a Sherman tank clearing Japanese infiltrators from their positions at dawn on Bougainville in March 1944.

AIR FORCE MUSEUM OF NEW ZEALAND, ALB87351048

RIGHT | A gruesome marker beside King's Way on Bougainville. This photograph was taken in March 1945.

AIR FORCE MUSEUM OF NEW ZEALAND, PR5459

ABOVE | An Australian mortar team in action in March 1945 near the Numa Numa Trail, which ran from the east coast over the central mountains of Bougainville to Torokina on the western coast.

AIR FORCE MUSEUM OF NEW ZEALAND, PR5423

ABOVE | New Zealand infantry soldiers about to fire a mortar on Green Island in February 1944. On the left is Fred Mazengarb from Gisborne. His companion, from Temuka, is William, but his surname is unknown.

ROYAL NEW ZEALAND RETURNED AND SERVICES' ASSOCIATION, NATIONAL OFFICE

RIGHT | Shirts were optional at mealtimes in a tropical climate.

ENGINEER CORPS MEMORIAL CENTRE, BISHOP 073

ABOVE | A task those in Italy knew so well — digging out a vehicle stuck in the mud.

ENGINEER CORPS MEMORIAL CENTRE, BISHOP 032

ABOVE | New Zealand Engineers in their workshop.

ENGINEER CORPS MEMORIAL CENTRE, BISHOP 049

LEFT | Landing vital stores.

ENGINEER CORPS MEMORIAL CENTRE, IMG 0067

RIGHT | Members of 20 Field Company take the opportunity to bathe at the Uzama River bridge on Vella Lavella in the Solomons.

ENGINEER CORPS MEMORIAL CENTRE, ECMC 5886

LEFT | Sappers of 20 Field Company construct a jetty at Maravari on Vella Lavella.

ENGINEER CORPS MEMORIAL CENTRE, ECMC 5888

RIGHT | A New Zealand Engineer folding a boat raft used for crossing a tidal creek in New Caledonia.

ENGINEER CORPS MEMORIAL CENTRE, ECMC 58897

BELOW | Dealing with sacks of mail from New Zealand.

ENGINEER CORPS MEMORIAL CENTRE, NZE 1

ABOVE | New Zealand Engineers' plant equipment being unloaded on Nissan Island.

ENGINEER CORPS MEMORIAL CENTRE, WH 495

BELOW | New Zealand sappers learn to bake bread in the Pacific, in ovens they made themselves.

ENGINEER CORPS MEMORIAL CENTRE, BISHOP 056

LEFT | Sappers construct a model of Green Island for planning purposes.

ENGINEER CORPS MEMORIAL CENTRE, BISHOP 044

ABOVE | Japanese soldiers being repatriated after the war line up to receive a course of inoculations before their departure.

NATIONAL ARMY MUSEUM, 1998-2799

RIGHT | Japanese prisoners work under the less than watchful eyes of their Australian guards, probably on Bougainville in 1945.

AIR FORCE MUSEUM OF NEW ZEALAND, ALB02316150

ABOVE | A 25-pounder gun being moved into position before crossing the Sangro River in November 1943.

NATIONAL ARMY MUSEUM, 1991-980

| CHAPTER 9 |

The Italian Campaign

The Italian campaign of 1943–45 was, after North Africa, the second most significant New Zealand land commitment of the Second World War. For 18 months 2 NZEF and 2 New Zealand Division fought against some of the best German divisions of the Wehrmacht in terrain that greatly favoured those defending it. In Italy the New Zealanders experienced considerable hardships and several defeats before finally emerging triumphant in the last months of the war.

For the Allies there was an undeniable logic to invading Italy after the successful conclusion to the North African campaign. Italy was weakened by this defeat and disillusioned with the war. An invasion could knock Italy out of the war and would open up a second front that would appease the Russians, who had been pressuring their fellow combatants to do more. The resources for the campaign were already on hand: the British Eighth Army and a United States army, along with landing craft and shipping used in Operation 'Torch', the Allied invasion of French North Africa that began in November 1942 and ended, successfully, in May 1943. Capturing Italy would provide airfields close to Eastern Europe from which to bomb strategic targets, especially the Ploesti oilfields in Romania. There was also the need to maintain momentum after the German–Italian defeat in North Africa. For these reasons British and American forces landed on Sicily on 10 July 1943 and, after that island was secured in mid-August, moved on to invade the Italian mainland on 3 September.

It was a controversial decision and one the Americans agreed to only reluctantly. For the United States, the cross-Channel invasion of France was what really mattered; the Italian campaign would divert valuable resources from this objective. The British, however, were enthusiastic, especially as senior Italian figures were secretly negotiating with the Allies for their country's surrender. Prime Minister Winston Churchill, in particular, anticipated an easy and quick victory in Italy, which could offer further opportunities in the Balkans and central Europe. Primarily through his political skills, the British persuaded the unenthusiastic Americans to commit to the Italian campaign. Even then, the American political and military leaders did not want the advance to continue beyond Rome. As Antony Beevor has commented, 'The whole Italian campaign would be dogged by misconceptions and wishful thinking.'[1]

The initial landing in Italy indicated just how arduous and costly the campaign would be. The British Eighth Army, commanded by General Bernard Montgomery, landed unopposed at Reggio di Calabria, on Italy's toe, and began a steady advance. For the American–British Fifth Army, commanded by Lieutenant-General Mark Clark, the landing further north at Salerno was a completely different experience. The German commander, the talented Generalfeldmarschall Albert Kesselring, managed to split the American and British elements of the Fifth Army and almost drove them into the sea. Only the Allied naval gunfire support and air superiority saved the army from a catastrophe. As one British observer remarked, 'What we saw was ineptitude and cowardice spreading down from the command, and this resulted in chaos.'[2] The Germans eventually withdrew to the Gustav Line in the centre of Italy, having inflicted 9000 casualties on the Fifth Army. It was not a great start to a campaign that was meant to be easy. In the words of Max Hastings, 'the stage was set for eighteen months of slow and costly fighting in some of the most unyielding country in Europe'.[3]

THE MEN OF 2 New Zealand Division arrived in this 'unyielding country' in mid-October 1943. Because time was needed to build up and concentrate the division, it was not ready for action until a month later. The New Zealanders' first task was to cross the Sangro River and push on to the Gustav Line position near Orsogna. The division entered the front line on 14 November and began edging up to the southern bank of the Sangro. At 2.45 a.m. on 28 November the New Zealanders crossed the river with two infantry brigades and captured the steep hills beyond it. They then pushed onto the town of Castel Frentano, which they captured on 2 December. For a moment it looked as if the Eighth Army might break through the Gustav Line in the east, but the Germans quickly reinforced the line at the critical town of Orsogna. This would be the New Zealanders' next objective.

On 3 December, the division attacked and captured Orsogna but was driven out by a spirited German counterattack. This was the first occasion on which New Zealand soldiers came up against the new and superior German Tiger tank. Heavy, and armed with an 88-millimetre gun, this tank was vastly superior to the M4 Sherman used by the New Zealanders. A Sherman's gun could not penetrate the thick front armour of a Tiger. Over the next three weeks the New Zealanders made several more attempts to capture Orsogna. The attempt on 7 December made little progress; a two-day battle that began on 15 December took some ground but also failed. As a stalemate set in and the weather deteriorated the New Zealanders' morale plummeted. The conditions they faced were reminiscent of the worst

RIGHT | Gunners carry ammunition uphill for the Sangro River crossing.

NATIONAL ARMY MUSEUM, 1991-980

LEFT | Evidence of a New Zealand war crime, the body of a German soldier lies near the banks of the Sangro River. He had been captured by Private Ivan Rankin of 26 Battalion but was later shot in cold blood by another soldier. 'Murdered German POW' was Rankin's original caption.

NATIONAL ARMY MUSEUM, 2016-29-1

RIGHT | The 5 Brigade Signals Section at Orsogna.

NATIONAL ARMY MUSEUM, 2019-3-1

RIGHT | The Allied cemetery near Orsogna. Orsogna was 2 New Zealand Division's first setback in Italy, at a cost of over 1600 casualties.

NATIONAL ARMY MUSEUM, 2011-380-9

LEFT | A soldier examines a Tiger tank, one of the most feared and most powerful tanks of the Second World War, destroyed by 2 New Zealand Division.

NATIONAL ARMY MUSEUM, 2011-380-7

BELOW | Gun drill in the Liri Valley. The entrance to the valley was overlooked by the Benedictine monastery of Monte Cassino.

NATIONAL ARMY MUSEUM, 1991-980

experienced on the Western Front in the First World War. These circumstances contributed to a minor mutiny when the men of one platoon refused to follow their officer into action in a Christmas Eve attack on Orsogna. As Howard Kippenberger wrote, 'Such a thing was unheard of in the Division and the C.O. was heartbroken.'[4] But this unprecedented event indicated that the men in the front line were reaching the limits of their endurance.

As the weather worsened, there were no further attempts on Orsogna. In mid-January 1944, the New Zealanders were withdrawn and moved to the other side of the peninsula to take part in the Cassino battles. No one was sorry to be leaving Orsogna. New Zealand's first actions in the Italian campaign mirrored the experience of the Allied landing in September. It had begun well, but had disintegrated into frontal assaults on a heavily defended position that were all costly failures. In just two months of fighting 2 New Zealand Division had suffered more than 1600 casualties, mostly from the infantry battalions.[5]

THE SITUATION WAS no better at Cassino, where the New Zealanders experienced the nadir of their Italian campaign. The four battles of Monte Cassino involved six months of attritional warfare, attacking the German defenders at the strongest part of the Gustav Line. The landing at Anzio on 22 January 1944, during the First Battle of Cassino, was a daring initiative with some prospects of success, but it soon collapsed into a stalemate with the potential to become a serious military defeat for the Allies.

When 2 New Zealand Division moved to the Cassino area in January it was transferred from Eighth Army control to be part of Clark's Fifth Army. Surprisingly, he was not pleased to have the New Zealanders under his command, regarding them as 'unwelcome intruders' and their commander, Freyberg, as a 'primadonna' who required 'kid gloves handling'.[6] The division was initially to have an exploitation role — that is, to press a defeated enemy withdrawing from the battlefield — for which, with its armoured brigade and thousands of vehicles, it was well suited. However, with the failure of the First Battle to break the Gustav Line, this role was changed to one of assault.

In early February the New Zealand Corps was formed, consisting of 2 New Zealand Division, 4 Indian Division and 78 British Division. Its task was to break into one of the strongest positions on the German front. As a prelude to the attack, Freyberg requested that the Benedictine monastery on Monte Cassino be destroyed by aerial bombing. This, which occurred on 15 February 1944, was one of the most controversial actions of the whole Italian campaign and the real tragedy was that it served no purpose. The Indians, who had initiated the request, were not informed that the bombing was going ahead and so were not ready to attack the monastery once it had been destroyed. Instead the Germans moved into the ruins and transformed them into defensive strongpoints. When the Indians attacked two days later they were easily driven off. Down on the plain an attack on the town of Cassino by two companies of 28 (Maori) Battalion also failed. Although the Māori infantry captured the town's railway station, no support could reach them. A smoke barrage provided little shelter from German mortar fire. In the mid-afternoon the Germans launched a counterattack under cover of the New Zealand smoke. With no anti-tank weapons and no communication back to the division, the soldiers had little option but to withdraw. The battalion had suffered 130 casualties and their withdrawal ruined any hopes of success.[7]

The Third Battle of Monte Cassino was a much bigger and longer operation. Delayed for three weeks by bad weather, the attack began on 15 March 1944 with a massive aerial and artillery bombardment that reduced Cassino to rubble. When the artillery barrage stopped, New Zealand infantry and tanks were sent in to clear the town. The German paratroopers who had survived the bombardment emerged from their shelters and put up a tenacious defence. The fighting, from ruin to ruin, lasted for nine days with grim conditions similar to those experienced on the Eastern Front. Cassino had, in Ian McGibbon's words, become 'a mini-Stalingrad'.[8] The New Zealanders captured most of Cassino but not the Continental Hotel and other strongholds on the western edge of the town. The Indians, struggling in the heights above, took some ground but failed to capture the ruins of the monastery. On 23 March Freyberg decided that neither the New Zealanders nor the Indians could do any more and he called off the offensive. The ground taken was consolidated and another deadlock set. This third battle had killed 343 New Zealanders. The second and third Cassino battles were 'a costly failure which ranks alongside Passchendaele as among New Zealand's worst disasters'.[9]

It was in mid-May 1944, during the Fourth Battle of Cassino, that the Allies were able to concentrate overwhelming force against the German defenders in Operation 'Diadem'. Using 18 infantry and three armoured divisions with an additional three armoured brigades against 13 German divisions, and with total air supremacy, the Allies finally smashed through the Gustav Line. The French Expeditionary Corps first penetrated beyond the Gustav Line in the mountains, and the Polish Corps carried out a series of costly frontal assaults that finally captured the ruins of the Monte Cassino monastery.

ABOVE | New Zealand gunners using drag ropes to move their 25-pounder into position at Cassino.

NATIONAL ARMY MUSEUM, 1991-980

LEFT | An anti-aircraft gun in position at Cassino. This was part of 14 Light Anti-Aircraft Regiment, New Zealand Artillery.

NATIONAL ARMY MUSEUM, 1987-1930-3

ABOVE | Bombs fall on the monastery at Monte Cassino on 15 February 1944. The bombing actually assisted the German defenders, who occupied the ruins before 4 Indian Division was ready to attack.

NATIONAL ARMY MUSEUM, 2019-3-1

BELOW | A humorous warning sign near Cassino.

NATIONAL ARMY MUSEUM, 2019-3-1

LEFT | New Zealand soldiers pose in the doorway of a bombed house in the town of Cassino.

NATIONAL ARMY MUSEUM, 1987-1930-3

ABOVE | New Zealand soldiers wash and bathe in the Liri River. The locals do not seem to mind.

NATIONAL ARMY MUSEUM, 2011-380-9

ON 4 JUNE ROME was taken after the Germans evacuated it and declared an open city. It was the first of the European and former enemy capital cities to be liberated, but victory celebrations were muted. Wanting to be known as the liberator of Rome, Clark had disobeyed orders in capturing the city, which had allowed an entire German army to escape. The Germans, under Kesselring, retired in good order to form a new defensive position, the Gothic Line. This meant still more hard fighting ahead for the Allies in Italy. Although Operation 'Diadem' was the largest American–British land operation of war to date, it was quickly overshadowed. Two days after the capture of Rome, Operation 'Overlord', the D-Day landing in France, was launched on 6 June. All eyes now turned to France. To Churchill's fury, the Americans insisted on withdrawing six divisions from Italy to fight in France,[10] and the Italian campaign took second place.

From July to December 1944 there was a series of battles along the Gothic Line. The appalling winter conditions, the tough terrain, especially the rivers and mountains, and the resolute, skilled enemy well dug in to defensive positions made the later part of the Italian campaign just as difficult as its first six months. From 22 July 1944, 2 New Zealand Division was involved in the difficult battles for Florence. The city was not taken until 4 August; more than 1000 New Zealanders became casualties in the fighting.[11] Then came a number of river crossings in the north of Italy to reach and capture the town of Faenza on 14 December. By Christmas that year the New Zealanders had reached the Senio River, where they took up defensive positions and settled in to face another freezing Italian winter.

It was not until 8 April 1945 that the New Zealand Division, which had been reorganised in early 1945 to create an additional infantry brigade, crossed the Senio in a set-piece attack with excellent close air support and artillery. It made rapid progress. Germany was now on the brink of defeat with Allied forces across the Rhine and the Soviet armies pressing in from the east. After the New Zealanders crossed the Senio came the Santerno (11–12 April) and the Sillaro a few days later. Because strong German opposition prevented an easy crossing of the Gaiana River, the New Zealanders had to prepare another set-piece battle; it would be their last. They were up against their old foes, six battalions of paratroopers supported by Panther tanks with 26 Panzer Division in reserve. The attack began on the night of 19 April with a heavy artillery barrage followed by a devastating flame attack using Wasp and Crocodile tanks. The infantry advanced immediately after the flames ceased. 'Progress was so rapid that it was initially thought the enemy must have gone, but the morning light revealed the carnage caused.'[12]

The division advanced rapidly, crossing the Reno River and reaching the formidable Po on 23 April. This they crossed two days later, on the thirtieth anniversary of the Gallipoli landing. The New Zealanders then pushed on to Venice and beyond to reach Padua on 28 April. They then received orders to race for Trieste to prevent it being captured by Yugoslav partisans who wanted to retain the Italian city for their country. On 1 May the division crossed the Isonzo River, where so many Italians and Austrians had perished in the First World War, and reached Montefalcone the next day. They arrived in Trieste about 4 p.m. on 2 May, the same day the Yugoslav partisans entered the city. Pockets of German resistance readily surrendered to the New Zealanders. In the three weeks since embarking on its new offensive, 2 New Zealand Division, as the spearhead of the Eighth Army, had advanced more than 320 kilometres and destroyed any opposition it encountered. It was an impressive feat of arms.

It is not well known but Trieste was the site of the only German concentration camp in Italy, set up in a disused rice factory at San Sabba. Since September 1943 it had held 25,000 victims, 5000 of whom had been killed by gas vans or shot at nearby firing ranges. As the New Zealanders approached the city the SS guards, in an effort to conceal their crimes, released the remaining inmates and destroyed the factory.[13]

In his history of the Second World War, Max Hastings aptly subtitled his chapter on Italy 'High hopes, sour fruits'.[14] After failing to fulfil initial expectations, the campaign dragged on for nearly two years as the Allies slowly advanced up the peninsula. The campaign did tie down some 15 per cent of German ground forces and result in more than half a million German casualties, double the Allied total.[15] It also took Italy out of the war, but this would have happened without the invasion. There is no doubt, though, that the Allies' presence prevented Italy from turning to communism after the war and contributed to the downfall of Nazi Germany. The New Zealanders played their part in a campaign that had to use every resource available to defeat a determined enemy. That they were not always used wisely or well is beyond dispute, but their efforts nonetheless contributed to the final outcome. Just over 2000 New Zealanders lie in Italian graves; another 7000 were wounded.[16] For those who survived the ordeal, Italy had been a brutal, bitter experience and they were immensely relieved when it was over.

ABOVE | 5 Brigade Signals Centre at Cassino. The slit trench looks hastily dug and offers little protection.

NATIONAL ARMY MUSEUM, 2019-3-1

BELOW | A close-up view of the ruins of Cassino with Castle Hill in the background.

OZ TURLEY, 6TH FIELD REGIMENT, FROM TONY GOODWIN

ABOVE | General Freyberg and Brigadier Keith Stewart at Cassino. Freyberg was under immense pressure during the Cassino operations, as well as having to cope with the knowledge that his only son, Paul, was missing in action. (It was later learnt that he survived.) Stewart, later chief of the general staff, was commanding 5 Brigade.

NATIONAL ARMY MUSEUM, 2019-3-1

ABOVE | Looking down on what remains of Cassino from Monastery Hill. Vehicles can be seen moving along Highway 6.

OZ TURLEY, 6TH FIELD REGIMENT, FROM TONY GOODWIN

RIGHT | The original graves of New Zealand soldiers killed at Cassino. The remains were later moved to the nearby Commonwealth War Graves cemetery.

NATIONAL ARMY MUSEUM, 1992-164

ABOVE | At Cassino General Freyberg, far right, contemplates the cost of the worst months of the war.

ROYAL NEW ZEALAND RETURNED AND SERVICES' ASSOCIATION, NATIONAL OFFICE

BELOW | Churchill tanks on the move loaded with New Zealand soldiers.

NATIONAL ARMY MUSEUM, 2007-693

ABOVE | One of the Germans' most feared weapons: a 21-centimetre Nebelwerfer 42 rocket launcher, near Cassino in early 1944.

ROYAL NEW ZEALAND RETURNED AND SERVICES' ASSOCIATION, NATIONAL OFFICE

LEFT | Members of the Medical Corps on cookhouse fatigue, peeling the spuds.

NATIONAL ARMY MUSEUM, 1990-1198

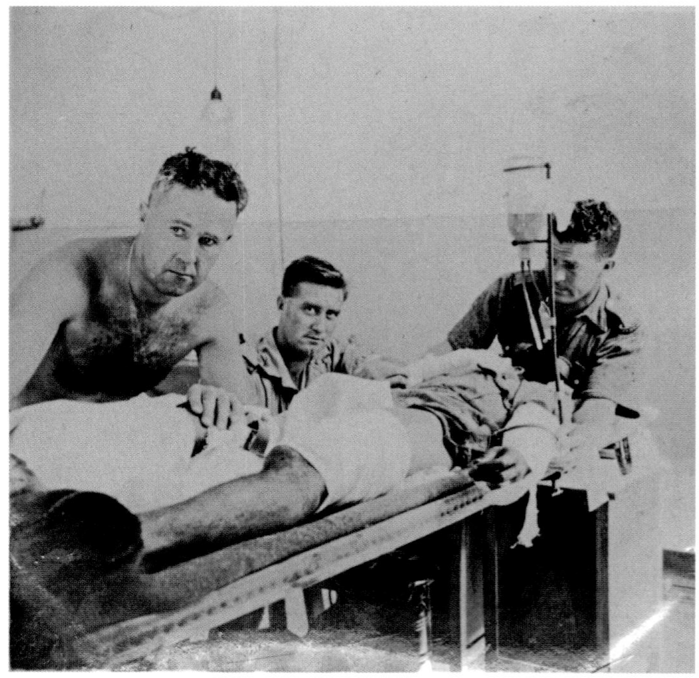

ABOVE | A wounded New Zealand soldier is given a blood transfusion.

NATIONAL ARMY MUSEUM, 1990-1198

LEFT | Soldiers attend a lecture on malaria at a New Zealand field hospital. The disease was prevalent in Italy, especially in the south.

NATIONAL ARMY MUSEUM, 1990-1198

BELOW | New Zealand sappers clear a minefield in Italy. Mines and booby traps were a constant hazard.

ENGINEERS CORPS MEMORIAL CENTRE, RNZE 33

ABOVE | New Zealand Engineers in Italy take heed of the warning.

ENGINEERS CORPS MEMORIAL CENTRE, RNZE 1944

RIGHT | Keeping New Zealanders in touch was essential during the war. Here Lieutenant Bill Washbourn of the Army Service Corps records a radio message to be sent home to Greymouth. Archibald Curry of the New Zealand Broadcasting Unit holds the microphone.

NATIONAL ARMY MUSEUM, 1992-2316

LEFT | A 25-pounder gun crew enjoy the Italian summer heat.

NATIONAL ARMY MUSEUM, 2007-693

RIGHT | In 1944 New Zealand soldiers welcome a change to summer uniform.

NATIONAL ARMY MUSEUM, 2011-380-9

LEFT | New Zealand soldiers take the opportunity to wash and dry their clothes.

NATIONAL ARMY MUSEUM, 2011-380-7

ABOVE | New Zealand soldiers pose at St Peter's Basilica in Rome.

NATIONAL ARMY MUSEUM, 2011-224-1

BELOW | With the Colloseum in the background, a truck crowded with New Zealand soldiers takes in the sights of Rome in June 1944.

NATIONAL ARMY MUSEUM, 2011-380-7

ABOVE | Two Tuis of the Women's Auxiliary Army Corps enjoy their first day in Rome.

NATIONAL ARMY MUSEUM, 2010-287-1

LEFT | New Zealand soldiers with Italian children at Rimini on Italy's Adriatic Coast.

NATIONAL ARMY MUSEUM, 1992-164

BELOW | A soldier named Reg poses with two Italian women.

NATIONAL ARMY MUSEUM, 2011-380-9

LEFT | New Zealand soldiers respected Italian women for their fortitude, work ethic and resilience. There were romantic liaisons and marriages.

NATIONAL ARMY MUSEUM, 1989-431

RIGHT | A fascist firing squad prepares to execute an Italian partisan.

NATIONAL ARMY MUSEUM, 1989-431

ABOVE | New Zealand soldiers examine a damaged Panther tank, which has lost its right track. Like the Tiger, the Panther had few vulnerabilities but its tracks could not be armoured.

NATIONAL ARMY MUSEUM, 2017-640-1

BELOW | Grim faces after their truck has rolled off the road.

NATIONAL ARMY MUSEUM, 2011-380-9

ABOVE | 'Keep coming, keep coming.' Loading an M4 Sherman tank onto a transporter. Carrying tanks like this saved wear on tracks, fuel and time.

NATIONAL ARMY MUSEUM, 2011-380-9

LEFT | German troops in the Italian hill country, ready to move. The Germans in Italy fought hard until the end of the war. Although outnumbered and lacking air cover, they were tenacious in defending their positions.

WAIRARAPA ARCHIVE, 10-145-007-12

RIGHT | German soldiers at a battlefield funeral.

NATIONAL ARMY MUSEUM, 2016-29-1

LEFT | New Zealand Prime Minister Peter Fraser, centre, visiting soldiers at Villa la Rotonda, Italy, in 1944. Generals Freyberg and Edward Puttick are in the back of the jeep.

NATIONAL ARMY MUSEUM, 2005-69

RIGHT | Another river to cross — New Zealand Engineers move a Bailey bridge into position.

NATIONAL ARMY MUSEUM, 1990-144

BELOW | A sign only a New Zealand Engineer would make. The Engineers bridged many Italian rivers during the campaign.

NATIONAL ARMY MUSEUM, 2007-693

ABOVE | New Zealand Engineers construct a ford as a temporary crossing point until a Bailey bridge can be built.

ENGINEERS CORPS MEMORIAL CENTRE, RNZE 20

ABOVE | Sergeant Ivan Rankin (left) and friend.

NATIONAL ARMY MUSEUM, 2016-29-1

ABOVE | Battle-hardened soldiers of 26 Infantry Battalion are ready to move again. This photograph was taken at the village of Barbiono in Italy's north-east.

NATIONAL ARMY MUSEUM, 2016-29-1

LEFT | Soldiers of 26 Infantry Battalion hitch a ride on an M4 Sherman tank of 20 Armoured Regiment. Both units were recruited from New Zealand's South Island.

NATIONAL ARMY MUSEUM, 2016-29-1

LEFT | A New Zealand convoy on the road to Florence in August 1944.

NATIONAL ARMY MUSEUM, 2011-380-7

BELOW | The first tank of 2 New Zealand Division receives a warm welcome from the local civilians upon entering the city of Florence in early August 1944.

NATIONAL ARMY MUSEUM, 2011-380-7

BELOW | New Zealand infantry advancing cautiously up to a wrecked bridge over the Lamone River on the outskirts of Faenza in December 1944.

NATIONAL ARMY MUSEUM, 2011-380-7

ABOVE | New Zealand soldiers watch their 1944 Christmas dinner being prepared.

NATIONAL ARMY MUSEUM, 1989-393

BELOW | A New Zealand soldier contemplates the ruins of Faenza in December 1944.

NATIONAL ARMY MUSEUM, 2011-380-7

ABOVE | A 'pee in the Po', everyone called it. Four soldiers take the opportunity to urinate in Italy's longest river.

NATIONAL ARMY MUSEUM, 2009-1042

BELOW | Dead German soldiers beside a road in Italy.

NATIONAL ARMY MUSEUM, 2011-224-1

ABOVE | Wasp flame-thrower tanks, like this one taking a practice shot, were used with devastating effect when the New Zealanders crossed the Gaiana River in April 1945.

NATIONAL ARMY MUSEUM, 2009-1042

ABOVE | New Zealand soldiers were rushed to Trieste to prevent it being captured by Yugoslav partisans.

NATIONAL ARMY MUSEUM, 2016-29-1

RIGHT | 'Poor Tom', reads the caption on this photograph. Tom was 33-year-old Otago man Private Sydney James Campbell of 26 Battalion, killed in action on 2 May 1945. He was married with two sons. Campbell's body was later moved to the Udine War Cemetery.

NATIONAL ARMY MUSEUM, 2016-29-1

LEFT | The New Zealand flag flying over Trieste in May 1945.

NATIONAL ARMY MUSEUM, 2011-380-7

ABOVE | Locals and soldiers at Trieste take advantage of the warm weather.

NATIONAL ARMY MUSEUM, 1989-393

LEFT | The people of Trieste celebrate as Tito's Yugoslav partisans leave the city.

NATIONAL ARMY MUSEUM, 2018-222-4

ABOVE | The war is over. Three New Zealand soldiers enjoy a brief Mediterranean cruise.

NATIONAL ARMY MUSEUM, 2019-3-1

RIGHT | Graves of three New Zealand soldiers near Cassino. The soldiers buried here are, front from left: Donald Tapuke of 28 (Maori) Battalion and Aubrey Charles Emanuel of 26 Battalion; behind: James Heke of 28 (Maori Battalion), who was awarded the Military Medal for his gallantry at Takrouna in North Africa. All three soldiers were killed in the battles for Monte Cassino.

NATIONAL ARMY MUSEUM, 2011-380-9

ABOVE | A sign marks the limit of the Allies' advance in Italy.

NATIONAL ARMY MUSEUM, 2016-29-1

LEFT | New Zealand soldiers on an Italian train that will begin their long journey home.

NATIONAL ARMY MUSEUM, 2011-224-1

ABOVE | In this surreptitiously taken photograph a senior German general inspects the Red Cross store at Stalag 383. This image was in an album kept by Bombardier Robin Gray, who was captured on Crete.

NATIONAL ARMY MUSEUM, 1987-2305

| CHAPTER 10 |

Prisoners of War

During the Second World War more than 9000 New Zealanders became prisoners of war. The largest group, nearly 90 per cent,[1] was the 8348 members of 2 New Zealand Division captured during the fighting in Greece, Crete, North Africa and Italy. The debacles of Greece and Crete accounted for almost half of this number. Even in the 'victorious' Libyan battles of November–December 1941 a further 2042 men were taken prisoner.[2] The long chain of military disasters in which New Zealand shared, starting with the Greek campaign in April 1941 through to the turning point battle of Alam Halfa in September 1942, resulted in long casualty lists that included those who went 'into the bag'.

New Zealand airmen and sailors also became prisoners of war, although in far fewer numbers than their army counterparts. The involvement of New Zealanders in every great air campaign waged by the Allies took its toll, and among the casualties were 575 airmen who became prisoners of war; another 23 were interned in neutral countries.[3] Some 194 sailors, including those with the Royal Navy and Merchant Marine, became prisoners of war or were interned.[4]

MOST CAPTURED New Zealanders were prisoners of the Italians or the Germans, taken in North Africa or Europe. Of the few hundred who became prisoners of the Japanese, the majority were New Zealand civilians who had been working in Hong Kong, Malaya, Burma and Singapore when these places were overrun by Japanese forces. The exact number may never be known, but according to the official history 377 New Zealanders were held by the Japanese during the war. Of these 214 were civilians interned in the Pacific, 51 were members of the Merchant Navy, 52 were sailors, 53 were airmen and seven were soldiers of the NZEF.[5]

There is no doubt, though, about what these prisoners endured. They were starved, worked to exhaustion, beaten, kept in inadequate shelter and cruelly mistreated. For the Japanese, steeped in the warrior code of bushido, to be taken prisoner meant total disgrace. People who surrendered were believed to have failed in their duty to the sovereign and to the nation. Little wonder that the Japanese military treated prisoners with contempt and cared little for their welfare. Although there were few deaths, the suffering of these captives left an undercurrent of bitterness that lasted for decades. The Australian experience was particularly appalling: about a third of their prisoners of war, some 8000 men, died in captivity, most of them in Japanese camps.[6] Harrowing accounts of their experiences written by Australian authors became bestsellers and essential reading for a generation of New Zealanders.[7]Although never as harsh as life under the Japanese, imprisonment in a German or Italian camp was not easy. Many men went through long years of captivity where 'boredom, deprivation and hunger were their lot'.[8] In late 1944 and early 1945, when Soviet forces pressed in on Germany's eastern borders, many Allied prisoners, New Zealanders among them, had to struggle through a long forced march in freezing weather. Some were on the road for months and not all of them survived the ordeal.

REMARKABLY, JUST OVER 700 captured New Zealanders managed to escape and reach safety.[9] Many more made unsuccessful escape attempts, including Victoria Cross recipients Charles Upham and Leonard Trent. The latter took part in what was later called the 'Great Escape' from Stalag Luft III at Sagan (in what is now Poland) in March 1944, but was recaptured almost as soon as he emerged from the tunnel.[10] He was perhaps lucky to be taken so early; 50 of the 76 men were later shot on Hitler's orders. Among those murdered were three New Zealand airmen: Porokoru (Johnny) Pohe, Arnold Christensen and John Williams.[11] Mick Shand, another New Zealander who took part in the Great Escape, was quickly recaptured and returned to the camp. After numerous unsuccessful escape attempts, Upham finished his time at Colditz, the notorious camp for obstinate prisoners of war.[12] For most prisoners, though, their story was 'undramatic . . . Though such a theme does not lend itself to heroics, it contains some fine examples of courage and self-sacrifice; and daring, determination, and endurance.'[13]

No images of prisoners of Japan were found when compiling this book, and the reason for this is simple. Owning and using a camera would have attracted the severest of punishment from the Japanese guards and the risk was too extreme. Cameras were almost entirely banned in German and Italian camps. Even on special occasions such as theatre productions and sporting events the use of cameras and the development of the film were strictly controlled.[14] Despite this, there are many photographs of New Zealanders as prisoners of

LEFT | A song composed in an Italian camp captures the mood of many prisoners (prigionieri).

C.R. LEES ALBUM

RIGHT | The inhabitants of Room 20 at Poppi (PG 38).

C. R. LEES ALBUM

RIGHT AND ABOVE | The amateur band at Oflag VA, Weinsberg, led by John 'Chan' Cowling, whose signature appears top right. An oflag was an officers' camp. The Geneva Convention required separate camps for officers and other ranks.

C. R. LEES ALBUM

ABOVE | Described as 'a cheerful group at Stalag VIIIB', most of these prisoners of war were New Zealanders captured in Greece and Crete. They would face long years of captivity.

ENGINEER CORPS MEMORIAL CENTRE, 6339

RIGHT | A group of prisoners at Stalag VIIIB, at Teschen in Upper Silesia, pose with musical instruments. Only one man is named, Gunner John Bensemann of Nelson on the far right in the back row.

ENGINEER CORPS MEMORIAL CENTRE, 6340

war, mostly in German camps. Cameras were an essential tool for prisoners wanting to abscond. They could be used in the forging of travel documents and for taking images that could be used in planning escapes.

THE PRISONER OF WAR experience was an important part of New Zealand's involvement in the Second World War. For the first and only time in its history a considerable proportion of the nation's fighting force was held in enemy prison camps. For example, in May 1942, New Zealand had some 53,000 men on active service overseas and 6300 of them were captives. The 'preservation of their safety and welfare had become a task of considerable national importance'.[15] That significance was also recognised by the inclusion of a volume on prisoners of war in the impressive official history of New Zealand in the Second World War. Written by W. Wynne Mason, it was one of the first in the series to be published, and New Zealand remains the only warring nation to honour its prisoners in this way for their part in the war.

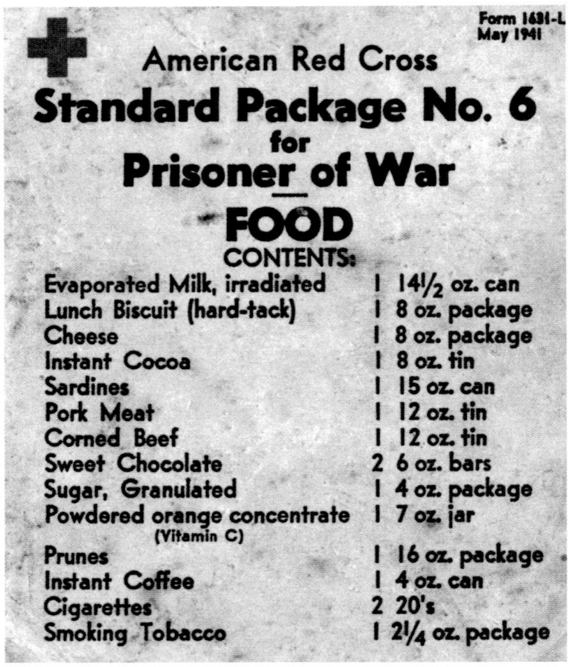

Form 1681-L
May 1941

American Red Cross
Standard Package No. 6
for
Prisoner of War
FOOD
CONTENTS:

Evaporated Milk, irradiated	1	$14\frac{1}{2}$ oz. can
Lunch Biscuit (hard-tack)	1	8 oz. package
Cheese	1	8 oz. package
Instant Cocoa	1	8 oz. tin
Sardines	1	15 oz. can
Pork Meat	1	12 oz. tin
Corned Beef	1	12 oz. tin
Sweet Chocolate	2	6 oz. bars
Sugar, Granulated	1	4 oz. package
Powdered orange concentrate (Vitamin C)	1	7 oz. jar
Prunes	1	16 oz. package
Instant Coffee	1	4 oz. can
Cigarettes	2	20's
Smoking Tobacco	1	$2\frac{1}{4}$ oz. package

LEFT | American Red Cross parcels were highly prized, especially for the generous tobacco allowance.

NATIONAL ARMY MUSEUM, 1987-2305

ABOVE | A happy day — distributing Red Cross parcels in Stalag 383. All prisoners acknowledged that these kept them alive.

NATIONAL ARMY MUSEUM, 1987-2305

RIGHT | A group of NCOs at Stalag 383 open Red Cross parcels ready to distribute their contents.

NATIONAL ARMY MUSEUM, 1992-1163

LEFT | War hero and a difficult prisoner to manage, Charles Upham would eventually be confined to Colditz, Oflag IVC, for numerous escape attempts. This photograph was taken in Syria in 1942. Upham was captured in July of that year.

MAJOR 'JO' JOHANSEN

BELOW | An ominous sentry box at Stalag 383. Note the prisoners' vegetable patch in the foreground.

NATIONAL ARMY MUSEUM, 1992-1163

RIGHT | In April 1943 Stalag 383, at Hohenfels in Bavaria, held more than 4000 prisoners, including 320 New Zealanders. Stalags were for other ranks: this one housed British NCOs.

NATIONAL ARMY MUSEUM, 1992-1163

RIGHT | Some New Zealand prisoners were forced to work for the Germans, in this case in a coal mine at Knurów in Poland. The man at the top left is Frank Muntz, who was captured on Crete on 1 June 1941 and worked at Knurów for almost a year. The 60 or so prisoners at the mine were forced to put in nine-hour days, seven days a week. Muntz feigned an injury to return to his original camp, Stalag VIIIB, at Teschen, in early 1943.

FRANK MUNTZ, POW CRETE

LEFT | Christmas dinner 1943 in Hut 146 at Stalag 383. This was the third Christmas in captivity for Arthur Dalton from Napier, who was a member of this hut. He was captured when 5 Brigade Headquarters was overrun in the 'Crusader' battle.

NATIONAL ARMY MUSEUM, 1992-1163

ABOVE | The New Zealand rugby team at Stalag 383, winner of the Empire Games held at the camp in 1943.

NATIONAL ARMY MUSEUM, 1987-2305

LEFT AND ABOVE | Stalag 383 also boasted a New Zealand soccer team and a cricket team, including white-coated umpires. Playing sport staved off boredom and helped to keep the men fit.

NATIONAL ARMY MUSEUM, 1987-2305, 2001-1015-1

ABOVE | A quick snap of a Red Cross inspection committee being shown around Stalag 383.

NATIONAL ARMY MUSEUM, 1992-1163

ABOVE | Sergeant James Kenneth Welch, who created this caricature in March 1943, was a display artist from Masterton. He was one of almost 2000 New Zealanders captured in the disastrous campaign in Greece in April 1941.

NATIONAL ARMY MUSEUM, 2001-1015-1

LEFT | Bleak surroundings and freezing weather make it hard for prisoner patients to raise a smile.

NATIONAL ARMY MUSEUM, 1987-2305

ABOVE | The strain shows on the faces of these prisoners outside their hut in Stalag 383, but there is still humour: the signboard reads 'NZ Tourist Bureau'.

NATIONAL ARMY MUSEUM, 1992-1163

RIGHT | Christmas dinner in Hut 24 of Stalag 383.

NATIONAL ARMY MUSEUM, 2001-1015-1

ABOVE | Prisoners at Stalag 383 take some exercise on a walking circuit in the camp.

NATIONAL ARMY MUSEUM, 1992-1163

LEFT | Two sergeants at Stalag 383 turn their hands to making wigs that were used in theatre productions.

NATIONAL ARMY MUSEUM, 1987-2304

ABOVE | The costumes in this production of *The Mikado* at Stalag 383 are of impressive quality.

NATIONAL ARMY MUSEUM, 1992-1163

RIGHT | Events such as this carnival at Stalag 383, with its competition and even a few party hats, provided entertainment, added to camp funds and gave prisoners something to look forward to.

NATIONAL ARMY MUSEUM, 1992-1163

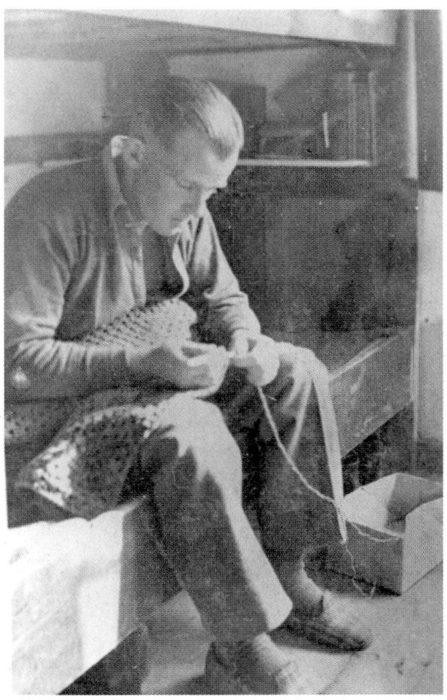

ABOVE | Surprising skills could come to the fore in camps. In Stalag 383, a New Zealander named Rogers is crocheting a rug.

NATIONAL ARMY MUSEUM, 1992-1163

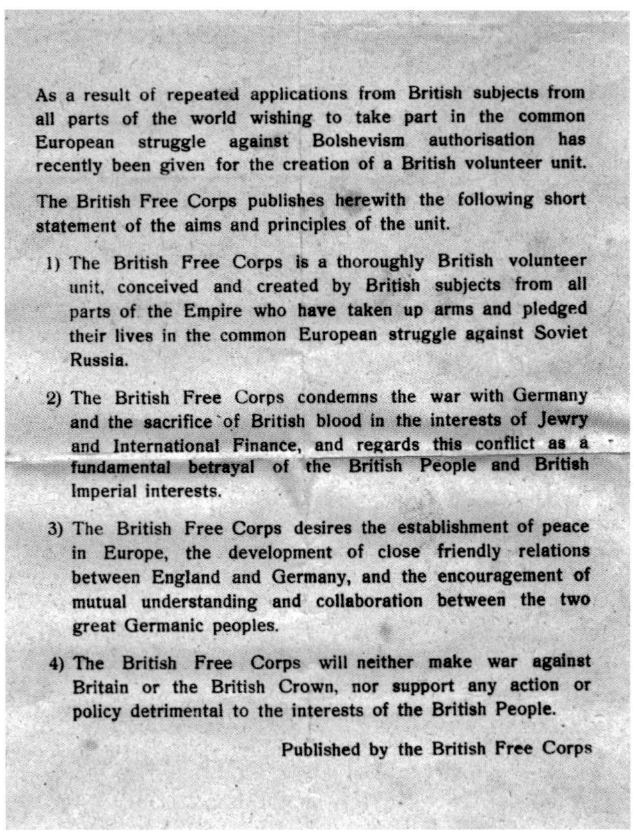

As a result of repeated applications from British subjects from all parts of the world wishing to take part in the common European struggle against Bolshevism authorisation has recently been given for the creation of a British volunteer unit.

The British Free Corps publishes herewith the following short statement of the aims and principles of the unit.

1) The British Free Corps is a thoroughly British volunteer unit, conceived and created by British subjects from all parts of the Empire who have taken up arms and pledged their lives in the common European struggle against Soviet Russia.

2) The British Free Corps condemns the war with Germany and the sacrifice of British blood in the interests of Jewry and International Finance, and regards this conflict as a fundamental betrayal of the British People and British Imperial interests.

3) The British Free Corps desires the establishment of peace in Europe, the development of close friendly relations between England and Germany, and the encouragement of mutual understanding and collaboration between the two great Germanic peoples.

4) The British Free Corps will neither make war against Britain or the British Crown, nor support any action or policy detrimental to the interests of the British People.

Published by the British Free Corps

LEFT | Sent by the British Free Corps, a Waffen-SS unit comprising British and dominion prisoners of war who had been recruited by Germany, this letter asks prisoners to join the Nazis in their fight against Bolshevism — that is, the Soviet Union. Only one New Zealander, British-born Roy Courlander, accepted their invitation. At his court martial Courlander maintained that he had joined the corps in order to spy and undermine its efforts, but he was sentenced to 15 years in prison, though he served only six before he was released.

NATIONAL ARMY MUSEUM, 2001-1015-1

ABOVE | A two-up table at Stalag 383. Note the packets of cigarettes used as gambling stakes.

NATIONAL ARMY MUSEUM, 1992-1163

RIGHT | Thanks to contraband radios like the one J. Paterson is listening to here, prisoners were well informed about the course of the war.

NATIONAL ARMY MUSEUM, 2001-1015-1

ABOVE | Prisoners watch Allied aircraft dominate the skies over Germany.

NATIONAL ARMY MUSEUM, 1992-1163

RIGHT | New Zealanders at Stalag VIIIB at Teschen, Germany, a large work camp that grew to hold 11,000 Commonwealth prisoners, including nearly 1000 Kiwis. One of the prisoners is wearing clogs.

NATIONAL ARMY MUSEUM, 2001-1015-1

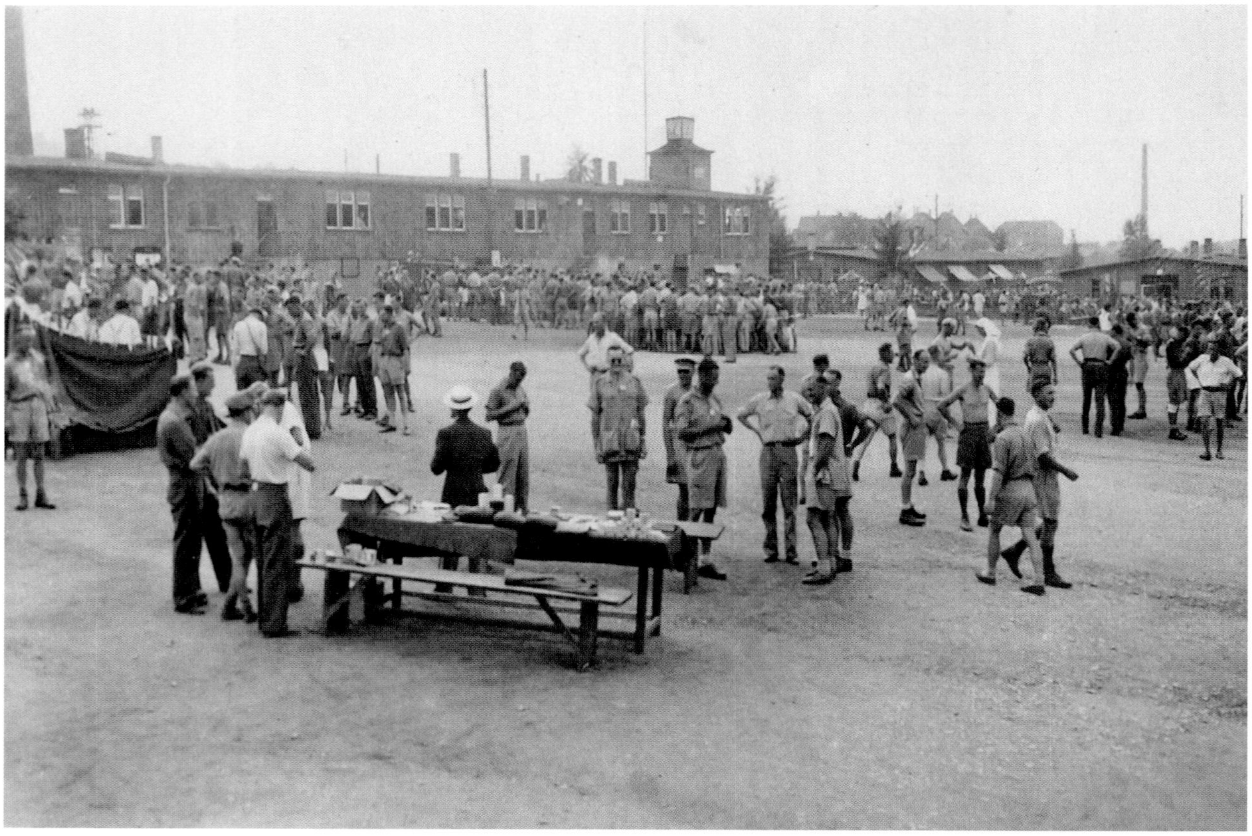

ABOVE | A 'Hampstead Heath' Fair Week at Oflag VA, Weinsberg, in July 1944 raised a significant amount of money for the YMCA.

C. R. LEES ALBUM

RIGHT | The Gatehouse Café was part of the Hampstead Heath Fair.

MAJOR 'JO' JOHANSEN

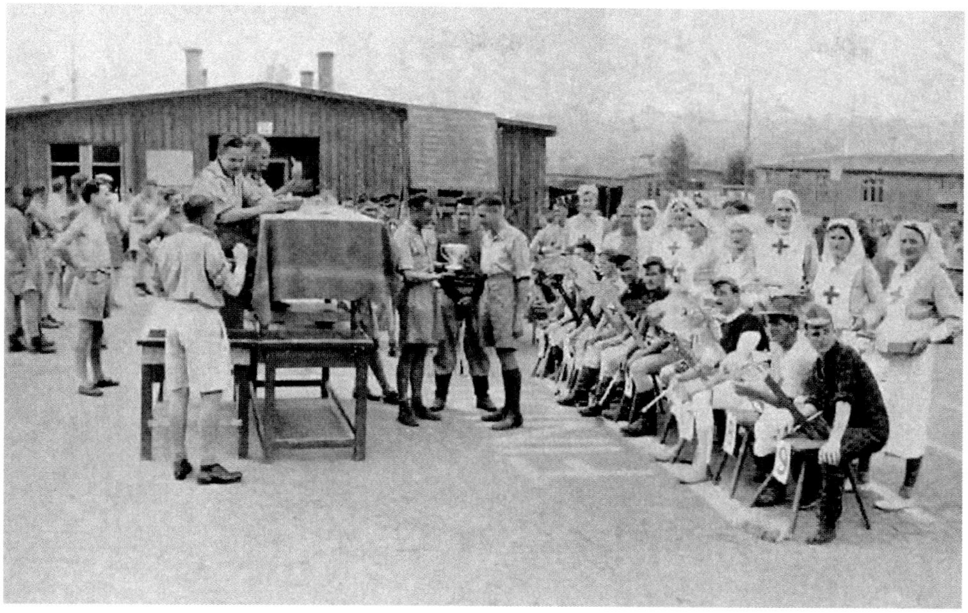

LEFT | Presenting sporting cups at Oflag VA. Note the 'nurses' in attendance.

MAJOR 'JO' JOHANSEN

BELOW | A sketch by E. T. Dobson of a roll call in the snow at Oflag XII B at Hadamer in Germany.

MAJOR 'JO' JOHANSEN

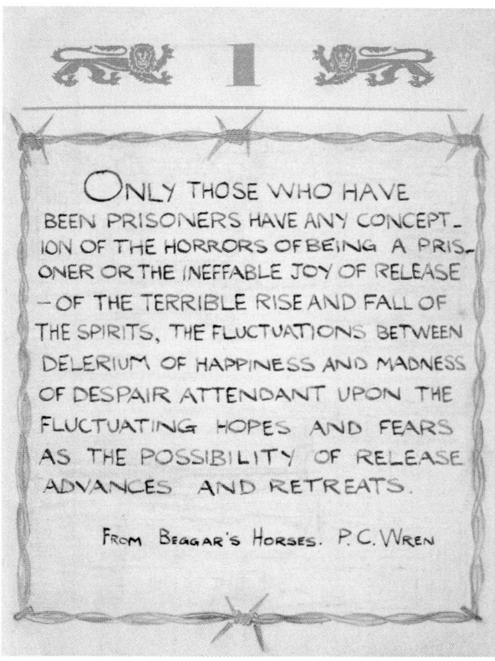

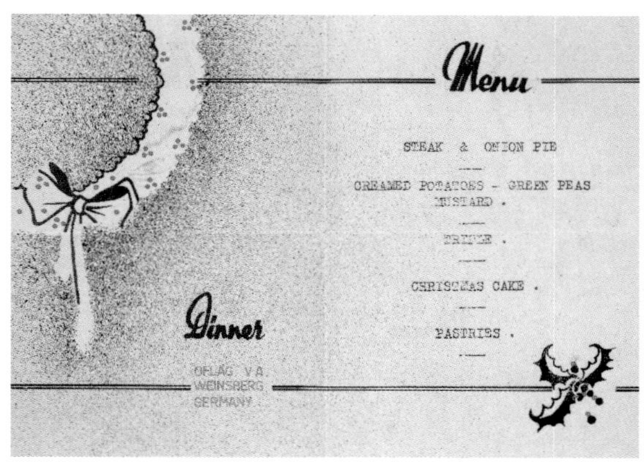

ABOVE | An extract from English writer P. C. Wren's 1934 book *Beggars' Horses* captures the emotional strain of being a prisoner of war. This was in a diary kept by Charles Ron Lees, who was captured at Sidi Rezegh in November 1941.

C. R. LEES DIARY

LEFT AND ABOVE | The 1944 Christmas dinner menu at Oflag VA.

C. R. LEES ALBUM

LEFT | Russian prisoners had it much worse than their British or American counterparts. The Soviet Union had not signed or acknowledged the Geneva Convention and, as a result, their treatment by the Germans was brutal. They were kept on starvation rations and worked hard, often to the point of death.

NATIONAL ARMY MUSEUM, 2001-1015-1

RIGHT | Liberation offered these desperate Russian prisoners little comfort since the Soviet authorities regarded them as no better than traitors and punished them accordingly.

NATIONAL ARMY MUSEUM, 2001-1015-1

LEFT | In the last years of the war, prison camp guards were as hungry as the prisoners they guarded. Here German guards rummage among discarded Red Cross cans searching for the remains of any food.

NATIONAL ARMY MUSEUM, 2001-1015-1

RIGHT | Former Stalag 383 prisoners on their way to Nuremberg airfield for a flight to Britain.

NATIONAL ARMY MUSEUM, 1992-1163

BELOW | The US Third Army, commanded by General George Patton, arrives to liberate Stalag 383.

NATIONAL ARMY MUSEUM, 1992-1163

ABOVE | Three naval ratings plus plenty of seagulls at the Mokohinau Island radar station sometime in 1942.

NATIONAL MUSEUM OF THE ROYAL NEW ZEALAND NAVY, OC171

| CHAPTER 11 |

The Home Front and Jayforce

The Second World War was a prime example of what historians call 'total war', which is defined as 'military conflict in which nations mobilise all available resources in order to destroy another nation's ability to engage in war'.[1] Total war demands that combatants, including small nations like New Zealand, dedicate not only their military forces to victory, but their entire civilian population as well. The Second World War was the most decisive event of the twentieth century and the 'greatest and most terrible of all human experiences'.[2] It asked of New Zealand, one of the earliest entrants to the war, its largest ever national effort.

An indication of New Zealand's determination to make the maximum possible contribution was the early introduction of conscription. After the first surge of volunteering in September 1939, the numbers of men coming forward steadily declined and the government quickly accepted that conscription was needed. When it came into effect in July 1940, it applied to men aged between 19 and 45, whose names would be drawn in ballots. The first ballot marble drawn in September 1940 was for service in New Zealand's Territorial Force; the first ballot for overseas service was carried out in December that year. Although there was some opposition to conscription, most of the public applauded its introduction. Later in the war, when the country fully mobilised its economy for war production, New Zealanders could be conscripted to work in essential industries. In January 1942 government legislation decreed that all men aged 18 to 49 were now at disposal of the National Service Department and could be directed to serve. This was known as 'manpowering' and shortly after its introduction it was applied to women, although limited initially to those aged 20 and 21. As the labour shortage grew worse, women aged between 18 and 40 could also be enlisted, unless they had children under 16. By December 1942 essential industries employed one-third of the workforce and 150,000 women had been conscripted into them.[3]

Many men who were ineligible for war service through age, ill health or because they were employed in an essential industry still wanted to play an active role. The Home Guard, formed in August 1940, provided the answer. Many New Zealand women were keen to serve, too. In 1940 the Women's War Service Auxiliary (WWSA) was formed and reached a peak of 75,000 members in mid-1942.[4] The armed forces also needed women to serve. The Women's Auxiliary Air Force (WAAF), established in January 1941, was soon followed by the Women's Auxiliary Army Corps (WAAC) in the middle of that year and then the Women's Royal New Zealand Naval Service (WRNZS, known as Wrens). By mid-1943 there were some 8500 women in uniform, most as WAACs. Most performed their duties in New Zealand, but some women also served overseas.

German raiders had already threatened the southern Pacific by sinking ships and laying mines across New Zealand harbour entrances. Then the island of Nauru was bombarded by German auxiliary cruisers in December 1940. Blackout was instituted in coastal parts of the country in February 1941. After the bombing of Pearl Harbor in December that year, the possibility of a Japanese invasion began to seem very real. As Prime Minister Peter Fraser said in his New Year's message, 'For the first time in our country's history we are directly threatened by enemy attack. At such a time I am proud to say, the people are neither dismayed nor are they unaware of the perils and difficulties which may be in store for them.'[5] The shock of the surrender of Singapore to the Japanese military on 15 February 1942, and the bombing of Darwin in northern Australia four days later, added to the concern and the feeling of vulnerability. Although the New Zealand government knew that an invasion by the Japanese was most unlikely, it encouraged the public to prepare for it, launching 'an unprecedented mobilisation' of its armed forces and people.[6] All coastal defences were manned, the 43,000 men of the Territorial Force were mobilised, defence works were constructed everywhere, including air raid shelters, trenches, tank traps, beach defences, pill boxes and underground headquarters. New Zealand was placed on a war footing, with the nightly blackouts now enforced by air raid wardens. The tension eased somewhat when US forces arrived in June 1942.

New Zealand's entire economy, including its agricultural output, was mobilised for the war effort. Control of essential goods required strict rationing, including, for the first time in the country's history, food. For some necessary items, such as butter and petrol, rationing lasted until 1950.

ABOVE | The radar crew at Mokohinau Island enjoying some sun.

NATIONAL MUSEUM OF THE ROYAL NEW ZEALAND NAVY, ABF0018

RIGHT | Part of the New Zealand garrison on Norfolk Island with bags of oranges they have collected.

NATIONAL ARMY MUSEUM, 1991-2284

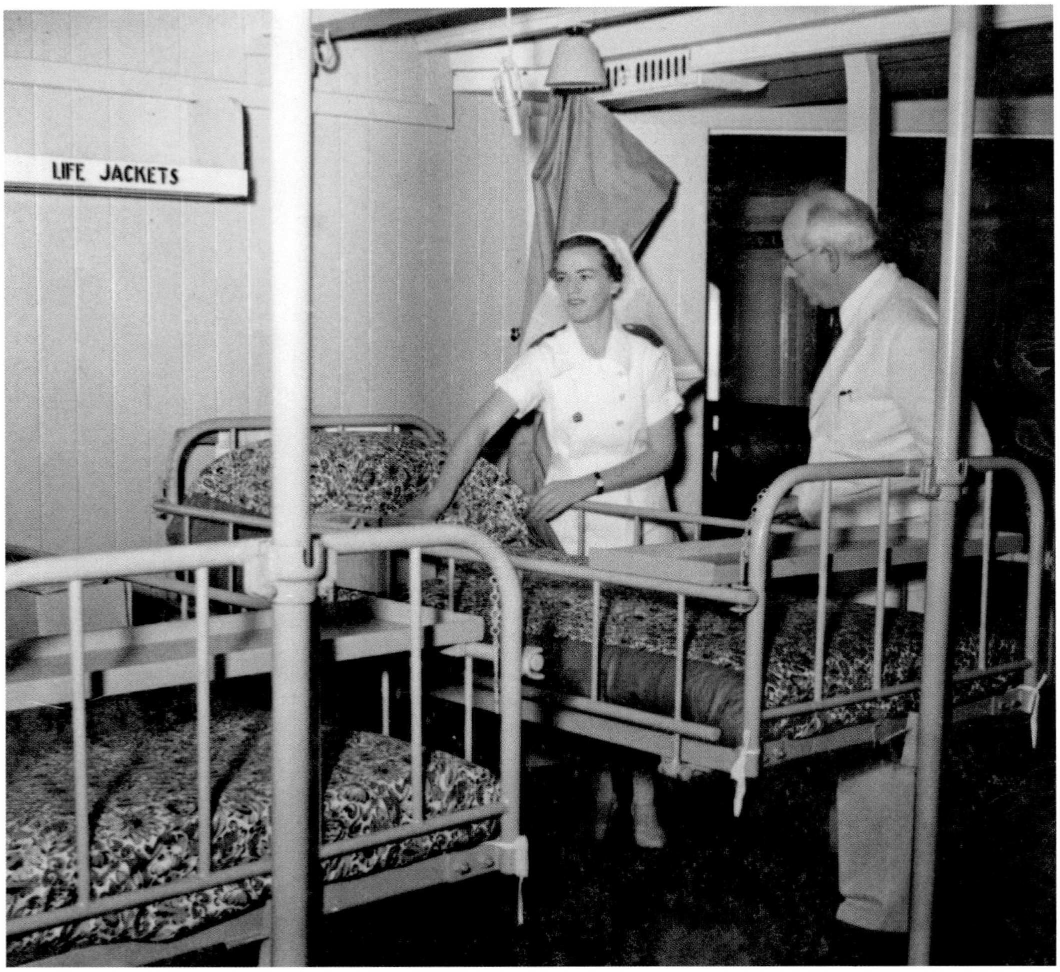

LEFT | Prime Minister Peter Fraser inspects what is most likely HMNZHS *Maunganui*, in which he is said to have taken a personal interest.

NATIONAL ARMY MUSEUM, 1990-223-2

RIGHT | The nursing staff of No. 1 New Zealand Hospital Ship (HMNZHS *Maunganui*) pose for a photograph at sea in May 1941.

NATIONAL ARMY MUSEUM, 1990-223-2

IN THE IMMEDIATE aftermath of the war it was HMNZS *Gambia* and *Achilles*, along with members of the Fleet Air Arm serving in the Royal Navy, that represented New Zealand's commitment to the occupation of a defeated Japan. Subsequently the government made available an infantry brigade and a fighter squadron as part of the Commonwealth Occupation Force. Later, the decision was made to raise, especially for this task, what became known as Jayforce, which was in Japan from mid-1946 through to the end of 1948. This was 'New Zealand's first contact with Asia on any substantial scale'.[7] Those who served in Jayforce may have modified their stereotypical perceptions of the Japanese people, but that country long remained a source of fear, suspicion and even hatred for many New Zealanders.

ABOVE | Part of the 6th Reinforcements in training at Trentham Camp in May 1941.

NATIONAL ARMY MUSEUM, 1994-3037

RIGHT | A group of hopeful women at RNZAF Station Wigram on 30 June 1941 filling in the forms required to join the Women's Auxiliary Air Force.

AIR FORCE MUSEUM OF NEW ZEALAND, WGF1511

ABOVE | 'Tigers on the tarmac. Harewood. April 1943' is the annotation on this photograph. The Southern Alps had a good dusting of snow that month.

AIR FORCE MUSEUM OF NEW ZEALAND, ALB10051162020

RIGHT | The composite minesweeper *T18*, HMNZS *Rimu*, at anchor in the Hauraki Gulf.

NATIONAL MUSEUM OF THE ROYAL NEW ZEALAND NAVY, OC87

RIGHT | This crowd has just watched Janet Fraser, wife of the prime minister, launch the minesweeper HMNZS *Awatere* at Evans Bay, Wellington, on 26 October 1942. The *Awatere* was just the second steel minesweeper built in New Zealand and the first warship built at Port Nicholson. Note the camera on top of the National Film Unit van.

NATIONAL MUSEUM OF THE ROYAL NEW ZEALAND NAVY, OC1271

ABOVE | United States First Lady Eleanor Roosevelt inspecting Wrens at HMNZS *Philomel* in early September 1943. Unlike their army counterparts, New Zealand Wrens did not serve overseas.

NATIONAL MUSEUM OF THE ROYAL NEW ZEALAND NAVY, 2013-70-777

LEFT | An early group of Wrens, photographed in 1942.

NATIONAL MUSEUM OF THE ROYAL NEW ZEALAND NAVY, APB0003

LEFT | In Christchurch's Cathedral Square, in front of the United Services Hotel, WAAFs, lined up in front of the wing of an Oxford trainer, take part in the 'Bonds for Bombers' week in March 1942. Behind them the WAAF Band drums up support.

AIR FORCE MUSEUM OF NEW ZEALAND, ALB041281067

BELOW | Chief Officer of the Women's Royal New Zealand Naval Service, Ruth Herrick, in 1942.

NATIONAL MUSEUM OF THE ROYAL NEW ZEALAND NAVY, APD0002

BELOW | WAAFs on parade dressing left, probably at RNZAF Station Wigram in 1942. The annotation on the image noted that the station warrant officer had been required to modify his language considerably when drilling the women.

AIR FORCE MUSEUM OF NEW ZEALAND, ALB041281005

ABOVE | A naval rating keeps a lonely vigil at the coast watching station at Oteranga Bay, Cape Terawhiti. This photograph was taken on 30 December 1942.

NATIONAL MUSEUM OF THE ROYAL NEW ZEALAND NAVY, OC400

ABOVE | Wrens practise their semaphore signalling on the beach at Matiu Somes Island in March 1943.

NATIONAL MUSEUM OF THE ROYAL NEW ZEALAND NAVY, OC433

RIGHT | At Matiu Somes Island in March 1943 a Wren sends a Morse code message using an Aldis signalling lamp.

NATIONAL MUSEUM OF THE ROYAL NEW ZEALAND NAVY, OC435

ABOVE | An air-to-air view of a formation of Fairey Gordons in flight over Christchurch.

AIR FORCE MUSEUM OF NEW ZEALAND, ALB841655080

BELOW | A Hudson of No. 1 Operational Training Unit in flight off New Zealand's coastline.

AIR FORCE MUSEUM OF NEW ZEALAND, OHG2956-54

ABOVE | WA2 Lowe working on a Tiger Moth in 1941.

AIR FORCE MUSEUM OF NEW ZEALAND, ALB041281013

BELOW | On the evening of 2 November 1943, the night log is updated at the Oteranga Bay Coast Watching Station, Wellington.

NATIONAL MUSEUM OF THE ROYAL NEW ZEALAND NAVY, OC418

ABOVE | The postal service delivers mail to the Oteranga Bay Coast Watching Station in November 1943.

NATIONAL MUSEUM OF THE ROYAL NEW ZEALAND NAVY, OC420

ABOVE | The motor launch *Q1183* patrols Wellington Harbour sometime in 1943.

NATIONAL MUSEUM OF THE ROYAL NEW ZEALAND NAVY, ABF0018

LEFT | Royal Marines from the aircraft carrier HMS *Indefatigable* on exercise in the Marlborough Sounds in December 1945.

NATIONAL MUSEUM OF THE ROYAL NEW ZEALAND NAVY, OC319

ABOVE | An Oxford trainer has flipped after landing downwind at RNZAF Station Nelson in 1942.

AIR FORCE MUSEUM OF NEW ZEALAND, ALB841655086

RIGHT | Corporal Reynolds uncovers a 500-pound UXB (unexploded bomb) while completing the No. 3 Bomb Disposal Course in August 1943.

NATIONAL MUSEUM OF THE ROYAL NEW ZEALAND NAVY, AUC0025

FAR RIGHT | And the bomb is destroyed.

NATIONAL MUSEUM OF THE ROYAL NEW ZEALAND NAVY, AUC0026

ABOVE | The Presentation of Sparks parade at Wigram in 1943 was a historic occasion because, for the first time, a WAAF member took the salute. Wing Officer Frances Kain, always known as Kitty, a dietitian from Dunedin, headed the WAAF from 1941 through to December 1943, initially as superintendent and later, when ranks were instituted, as wing officer. Kain resigned when pregnant with her second child. In 1949 she was awarded an OBE for her war service

AIR FORCE MUSEUM OF NEW ZEALAND, ALB041281139

RIGHT | A flight of WAAF members parades on the Wellington docks.

AIR FORCE MUSEUM OF NEW ZEALAND, ALB030131002

LEFT | A group of WAAFs attend the Officers' School of Instruction at RNZAF Station Levin in March 1943. This was the first course conducted with the aim of identifying potential WAAF officers and senior NCOs.

AIR FORCE MUSEUM OF NEW ZEALAND, ALB930322019

BELOW | Some of the Wrens at HMNZS *Philomel* who were responsible for sailing the commodore's barge. In this 1943 photograph Joy Bunting, on the right, is the only one named.

NATIONAL MUSEUM OF THE ROYAL NEW ZEALAND NAVY, OC290

LEFT | Four Wrens sail the commodore's barge in Waitematā Harbour.

NATIONAL MUSEUM OF THE ROYAL NEW ZEALAND NAVY, 2013-70-2

ABOVE | A depth charge fired by the modified Flower-class corvette HMNZS *Arabis* sails through the air during a November 1944 exercise in the Hauraki Gulf.

NATIONAL MUSEUM OF THE ROYAL NEW ZEALAND NAVY, OC600

ABOVE RIGHT | The charge explodes.

NATIONAL MUSEUM OF THE ROYAL NEW ZEALAND NAVY, OC139

RIGHT | During the same exercise, a depth charge from an unidentified ship sends up a wall of water.

NATIONAL MUSEUM OF THE ROYAL NEW ZEALAND NAVY, OC140

ABOVE | These WAAFs, who seem to be enjoying their work, are named, but the order is not specified, as Winifred Dee, Nellie Woods, Jean Murphy, Jean Hill and Marjorie Goffin. It is likely that the owner of the album, believed to be Daisy Elizabeth McCluggage, is also in the photograph.

AIR FORCE MUSEUM OF NEW ZEALAND, ALB9305023004

ABOVE | The Wrens who passed the Telegraphist Course at HMNZS *Tasman* in 1944. Back row, from left: Valerie Scott, Ruby Naish, Lynda Shirley, Jill Woodward, Yvonne Webb, Audrey Rambost, Ann Beetham, Riwia Apanui. Middle row: Jackie Bremner, Ursula Kay, Maureen Piper, Sylvia Denton, Betty Larmer, Yvonne Aronsen, Mary Newton. Front row: Mary Cholmondeley, Shirley Crawley, June Gasson, Joan Moffatt, Joyce Law, Elizabeth Smythe, Jean Baigent, Pauline Hansen. The instructor is Petty Officer George Gibbs.

NATIONAL MUSEUM OF THE ROYAL NEW ZEALAND NAVY, APD0030

ABOVE | HMNZS *Arabis* fires its anti-aircraft guns during an exercise in the Hauraki Gulf in November 1944.

NATIONAL MUSEUM OF THE ROYAL NEW ZEALAND NAVY, OC607

RIGHT | Three naval officers inspect torpedoes at the Devonport Dockyard in 1944. Lieutenant E. Savage, the base torpedo officer, is on the far right.

NATIONAL MUSEUM OF THE ROYAL NEW ZEALAND NAVY, OC660

ABOVE | These Palmerston North women are doing their bit for the national war effort in January 1941 in the evening after spending the day in their full-time jobs. They are, from left, unknown, Eileen Redwood, Rona Fenwick, Mrs Spike Moore, June Allen and Edna Spillman.

MANAWATŪ HERITAGE, 2012PA_IMCA-COOLSTORE_006564

RIGHT | Gathering hay was a difficult and exhausting task.

MANAWATŪ HERITAGE, 2012PA_IMCA-COOLSTORE_006565

RIGHT | New Zealand women did thousands of hours of voluntary work during the war years. The Polish Army League of Palmerston North, founded in December 1941, sent parcels and letters of friendship to Polish soldiers fighting with the British Army. From left are Mrs A. Stewart, Mrs L. E. Bale (President), Mrs T. K. Fraser (Treasurer) and founder and organising secretary Ann Jacques.

MANAWATŪ HERITAGE, 2011G_ELMAR-2-B8_005206

LEFT | Corporal Edna Spillman, shown gathering hay earlier, later served for four years with the New Zealand Army. She worked at the Railways Office arranging free railway passes for service personnel and their families.

MANAWATŪ HERITAGE, 2012PA_IMCA-COOLSTORE_006563

ABOVE | Jayforce soldiers parade before embarking for Japan.

NATIONAL ARMY MUSEUM, 1992-95

RIGHT | All aboard and ready to sail to Japan.

NATIONAL ARMY MUSEUM, 1992-95

ABOVE | Released at last, Australian prisoners of war from Wakayama, Japan, are taken onto landing craft. Part of their journey to freedom was on HMNZS *Gambia*.

NATIONAL MUSEUM OF THE ROYAL NEW ZEALAND NAVY, 2006.186.226.P40.1

RIGHT | Members of the crew of HMNZS *Gambia* in Tokyo in 1945. The rating on the right is Leslie Keith Subritzky from Waihopo, Northland.

NATIONAL MUSEUM OF THE ROYAL NEW ZEALAND NAVY, AAI0034

RIGHT | Early morning PT on the flight deck of British aircraft carrier HMS *Glory*, which left Auckland for Japan on 7 March 1946. This vessel carried 280 personnel of No. 14 Squadron RNZAF, who volunteered to serve in Japan and offer air support to the more than 4200 New Zealand troops who made up Jayforce.

AIR FORCE MUSEUM OF NEW ZEALAND, ALB89059B077

LEFT | Two days out from Auckland, a heavy swell and seasickness. Personnel on the flight deck of HMS *Glory*. Beside them are the aircraft taken to Japan, Corsairs and Dakotas.

AIR FORCE MUSEUM OF NEW ZEALAND, ALB89059B078

RIGHT | A New Zealand artillery unit drives through a Japanese city.

NATIONAL ARMY MUSEUM, 1989-431

ABOVE | 'Total destruction', reads the annotation on this photograph. A burnt-out tram and damaged building in Hiroshima in 1947.

AIR FORCE MUSEUM OF NEW ZEALAND, ALB890591B278

RIGHT | The devastation in Japan shocked the New Zealanders. This is what remained of Hiroshima after the dropping of the atomic bomb.

NATIONAL ARMY MUSEUM, 1986-2342

RIGHT | The ruins of the Kobe Steel Works at Chofu.

NATIONAL ARMY MUSEUM, 1986-2361

BELOW | Three New Zealand service personnel amid the destruction in Kure in 1947.

AIR FORCE MUSEUM OF NEW ZEALAND, ALB890591B172

ABOVE | Robert (Bob) Perkins sitting on the nose cone of a destroyed Japanese aircraft at Ozuki.

AIR FORCE MUSEUM OF NEW ZEALAND, ALB890591B198

RIGHT | The plight of a defeated nation. Japanese women and children search the shoreline for scraps of food.

NATIONAL ARMY MUSEUM, 1986-2342

RIGHT | In Kure in 1947 civilians and military personnel queue for black market food.

AIR FORCE MUSEUM OF NEW ZEALAND, ALB8890591B171

BELOW | Two New Zealand soldiers at Tokusa experience a Japanese winter.

NATIONAL ARMY MUSEUM, 1999-1688

ABOVE | New Zealand soldiers enjoying themselves at a Tokyo bar.

NATIONAL ARMY MUSEUM, 1986-2342

LEFT | A telling contrast — an armed patrol marching through a village and a kimono-clad woman beside a tree laden with cherry blossom.

NATIONAL ARMY MUSEUM, 2018-222-5

ABOVE | 'We Backed 'Em Up', reads the sign on this float, which features a working forge with young men, some in uniform, hammering and working with the metal. This was part of VJ Day celebrations in Palmerston North on 16 August 1945. The parade was followed by a community singsong at the band rotunda in the square.

MANAWATŪ HERITAGE, 2008P_WWT1A_WHF-1989

| CHAPTER 12 |

Coming Home

> *War is the greatest of all agents of change. It speeds up processes, wipes out minor distinctions, brings reality to the surface. Above all it brings home to the individual that he is not altogether an individual.*
> — *George Orwell,* The Lion and the Unicorn[1]

From mid-1945 and well into the following year New Zealand service personnel returned home to New Zealand to be 'demobbed' and return to their civilian lives. They came back as altered people to an altered country. As Nancy Taylor, the official historian of New Zealand's home front, perceptively concluded, 'To New Zealanders these [new] realities filtered through in treaties, trade and the texture of their everyday lives. As . . . Thomas Wolfe, American novelist, wrote, "You can't go home again"; the world and the people were changed.'[2] Adjusting to these changes was not always easy and it often took some time for returned service people to settle. Most managed the transition to civilian life, albeit not always smoothly, and got on with their lives as best they could. Compared with the experience of the First World War, the rehabilitation of veterans 'was a major success story'.[3] There were, however, many hidden and lasting scars of war and far too many 'silent casualties'.[4]

From a population of 1.6 million, some 140,000 New Zealand men and women served overseas, most of them, 104,000, in 2 NZEF. When the country's military mobilisation was at its highest, in July 1942, there were 154,549 men and women in the armed services and an additional 100,000 in the Home Guard. A total of 194,000 men served in New Zealand's armed forces; some 10,000 women donned military uniform. New Zealand also made a considerable economic commitment to the war — some 30 per cent of its national income overall, and this reached more than 50 per cent between 1942 and 1944.[5] And thousands of men and women were conscripted into essential industries, including farming.

A total of 11,625 New Zealanders lost their lives between 1939 and 1945. Although this was fewer than the more than 18,000 who did not return from the First World War, from a smaller population of just over one million, it was a high price to pay. Of note, too, was the fact that New Zealand's loss ratio per capita was the highest in the Commonwealth: 6684 per million compared with Britain's 5123 and Australia's 3232.[6] This reflected the combat roles most New Zealand units had in the war. The casualty list included more than 15,000 New Zealanders wounded in action and the 9000 who spent years in prison camps. To these numbers must be added the thousands of 'silent casualties' who continued to suffer when the fighting was over.

INITIALLY THERE WAS little doubt that, compared with the First World War, when accusations of futility and wasteful sacrifice by 'butchers and bunglers' emerged shortly after the armistice, the Second World War could be regarded as a 'good war'.[7] The general view was that it had been caused by a demonic man in charge of a powerful nation which fell under his spell. Adolf Hitler and Nazi Germany needed to be stopped and the only way to do this was by defeating them in battle. This war, then, was regarded as a moral crusade, one that had to be fought in order to end this evil and create a better world. This notion of a good or just war has been challenged over the years by revelations about Stalin's excesses in Eastern Europe, by the debate about the morality of strategic bombing, by the revelation of war crimes on both sides, and by the advent and course of the Cold War. It has also been challenged by historians from countries that were European colonial possessions at the time and who felt their people had been dragged into a war in which they had no interest or stake and for which they suffered heavily. There is some legitimacy to these claims but they do neglect a larger vision. Max Hastings' astute and eloquent assessment perhaps says it best:

> *Practically all those who participated, nations and individuals alike, made moral compromises. It is impossible to dignify the struggle as an unalloyed contest between good and evil, nor rationally to celebrate an experience, and even an outcome, which imposed such misery upon so many. Allied victory did not bring universal peace, prosperity, justice or freedom; it brought merely a portion of those things to some fraction of those who had taken part. All that seems certain is that Allied victory saved the world from a much worse fate that would have followed the triumph of Germany and Japan. With this knowledge, seekers after virtue and truth must be content.*[8]

ABOVE | A day after VE Day, which was 8 May 1945, a military band leads a parade around Palmerston North's Square to celebrate the end of the war in Europe.

MANAWATŪ HERITAGE, 2008_WWT7B_WHF-1997

LEFT | The war is finally over. General Yoshijirō Umezu, Chief of the Japanese Imperial Staff, signs the instrument of surrender on board USS *Missouri* in Tokyo Bay on 2 September 1945.

NATIONAL MUSEUM OF THE ROYAL NEW ZEALAND NAVY, AUC0002

LEFT | Celebrating VJ Day in Manners Street, Wellington, are, from left, Dorothy Grylls, Mabel Haase and Sister Eileen Santnett.

AIR FORCE MUSEUM OF NEW ZEALAND, ALB930322137

BELOW | Staff of the Navy Office in Wellington celebrate VJ Day on 15 August 1945.

NATIONAL MUSEUM OF THE ROYAL NEW ZEALAND NAVY, OC386

RIGHT | VJ Day celebrations on HMNZS *Achilles*. Raising a glass on the left is Chief Petty Officer W. S. Lambeth.

NATIONAL MUSEUM OF THE ROYAL NEW ZEALAND NAVY, AAF0169

ABOVE | The *Orion*, docked in London, is packed with New Zealand service personnel homeward bound.

NATIONAL ARMY MUSEUM, 2011.224.1

BELOW | 'I'm home.' A sailor of HMNZS *Gambia* is reunited with his family, but the little girl is unsure what to make of this stranger. This photograph was taken on 8 November 1945.

NATIONAL MUSEUM OF THE ROYAL NEW ZEALAND NAVY, OC296

ABOVE | A group of Tuis pleased to be homeward bound on the *Dominion Monarch* in 1945.

NATIONAL ARMY MUSEUM, 2010.287.1

BELOW | HMNZS *Gambia* arriving in New Zealand on 30 October 1945 from Japan. After being present for the Japanese surrender, *Gambia* transported prisoners of war freed from Japanese camps to Tokyo.

NATIONAL MUSEUM OF THE ROYAL NEW ZEALAND NAVY, 2006.186.226.P63.1

ABOVE | The Royal New Zealand Navy Contingent passes the saluting base during the Victory Parade held in London on 8 June 1946. King George VI and Queen Elizabeth are on the high dais. Below them, to the right, Winston Churchill stands second from the left.

NATIONAL MUSEUM OF THE ROYAL NEW ZEALAND NAVY, OC801

RIGHT | A lonely cemetery in Italy where three New Zealanders were buried.

NATIONAL ARMY MUSEUM, 1991.980

LEFT | In this photograph taken at the Beck Row Cemetery in England, the second and third headstones belong to Pilot Officer Arthur Patnell Pritchard, RNZAF, and his observer, RAF Sergeant Rupert James Bowden. Both men died when their No. 99 Squadron Wellington bomber crashed after take-off from Newmarket on 18 December 1940. Pritchard, from Whanganui, was just 20 years old. The cost of war is always the lives of young men and women.

AIR FORCE MUSEUM OF NEW ZEALAND, ALB051911013

ABOVE | An Anzac Day service at the National War Memorial, Wellington, in 2011. A Tri-Service vigil watches over the Tomb of the Unknown Warrior. We will remember them.

NEW ZEALAND DEFENCE FORCE, WELLINGTON

Notes

Introduction: New Zealand and the Second World War

1 Richard Holmes, *World War II in Photographs* (London: Carlton Books, 2000), 11.
2 Max Hastings, *All Hell Let Loose: The World at War 1939–45* (London: HarperPress, 2011), xv.
3 James Belich, *Paradise Reforged: A History of the New Zealanders: From the 1880s to the Year 2000* (Auckland: Allen Lane, 2001), 270.
4 Laurie Barber, *War Memorial: A Chronology of New Zealand and World War II* (Auckland: Heinemann Reed, 1989), 36.
5 Ian McGibbon, *New Zealand and the Second World War: The People, the Battles and the Legacy* (Auckland: Hodder Moa Beckett, 2004), 8.
6 John Crawford (ed.), *Kia Kaha: New Zealand in the Second World War* (Melbourne: Oxford University Press, 2002), 3.
7 These figures are from Crawford, *Kia Kaha*, 3–5.
8 Anna Rogers, *While You're Away: New Zealand Nurses at War 1899–1948* (Auckland: Auckland University Press, 2003), 200, 300–18.
9 Belich, *Paradise Reforged*, 270.
10 Crawford, *Kia Kaha*, 3.
11 Holmes, *World War II*, 7.
12 Major-General Sir Howard Kippenberger, KBE, CB, DSO, *Infantry Brigadier* (London: Oxford University Press, 1949), 1.
13 Kippenberger, *Infantry Brigadier*, 4.
14 Holmes, *World War II*, 9.
15 Michael King, *The Penguin History of New Zealand* (Auckland: Penguin, 2003), 408–09.

Chapter 1: At War Again

1 Ian McGibbon, *New Zealand and the Second World War: The People, the Battles and the Legacy* (Auckland: Hodder Moa Beckett, 2004), 14.
2 F. L. W. Wood, *The New Zealand People at War: Political and External Affairs* (Wellington: War History Branch, Department of Internal Affairs, 1958), 7.
3 McGibbon, *New Zealand and the Second World War*, 18.
4 Quoted in Wood, *New Zealand People at War*, 11. Various versions of Savage's speech were reported in the New Zealand newspapers. See, for example, the *New Zealand Herald*, 5 September 1939 and the *Evening Post*, 6 September 1939.
5 Wood, *New Zealand People at War*, 10.

Chapter 2: Early Days

1 Ian McGibbon, *New Zealand and the Second World War: The People, the Battles and the Legacy* (Auckland: Hodder Moa Beckett, 2004), 32, 33.
2 https://nzhistory.govt.nz/war/royal-new-zealand-navy/nz-division-royal-navy, 4, 5.
3 J. M. S. Ross, *Official History of New Zealand in the Second World War: Royal New Zealand Air Force.* (Wellington: War History Branch, Department of Internal Affairs, 1955), 33. See also W. David McIntyre, *New Zealand Prepares for War* (Christchurch: Canterbury University Press, 1988), 230.
4 McIntyre, *New Zealand Prepares for War*, 238, 246.
5 Ross, *Royal New Zealand Air Force*, vii.
6 The Governor-General of New Zealand to the Secretary of State for Dominion Affair, 13 September 1939, *Documents Relating to New Zealand's Participation in the Second World War 1939–45*, Vol. I, 21.
7 McIntyre, *New Zealand Prepares for War*, 230; Howard Kippenberger, 'The New Zealand Army', *Journal of the Royal United Services Institute* 102, no. 605 (1957): 71.
8 W. G. McClymont, *Official History of New Zealand in the Second World War 1939–45: To Greece* (Wellington: War History Branch, Department of Internal Affairs, 1959), 3; McIntyre, *New Zealand Prepares for War*, 247. The exact number of people who volunteered before conscription came into force in July 1940 was 59,644.
9 The High Commissioner for New Zealand (London) to the Prime Minister of New Zealand, 16 September 1939, *Documents Relating to New Zealand's Participation in the Second World War 1939–45*, Vol. I, 23.

10 Glyn Harper and Colin Richardson, *Acts of Valour: The History of the Victoria Cross and New Zealand* (Auckland: HarperCollins, 2016), 128–29.
11 Freyberg, letter to the Minister for Defence, 15 August 1940, *Documents Relating to New Zealand's Participation in the Second World War 1939–45*, Vol. I, 139.
12 Freyberg to the Minister of Defence, 23 February 1941, *Documents Relating to New Zealand's Participation in the Second World War 1939–45*, Vol. I, 207.
13 McIntyre, *New Zealand Prepares for War*, 247.
14 https://nzhistory.govt.nz/war/new-zealand-and-the-second-world-war-overview, 1.

Chapter 3: Greece and Crete

1 F. L. W. Wood, *The New Zealand People at War: Political and External Affairs* (Wellington: War History Branch, Department of Internal Affairs, 1959), 2.
2 Ian McGibbon, *New Zealand and the Second World War: The People, the Battles and the Legacy* (Auckland: Hodder Moa Beckett, 2004), 65.
3 Major-General Sir Howard Kippenberger, KBE, CB, DSO, *Infantry Brigadier* (London: Oxford University Press, 1949), 45.
4 Glyn Harper, *Kippenberger: An Inspired New Zealand Commander* (Auckland: HarperCollins, 1997), 68.
5 These figures are from W. G. McClymont, *Official History of New Zealand in the Second World War 1939–45: To Greece* (Wellington: War History Branch, Department of Internal Affairs, 1959), 486. Casualty figures are notoriously variable. Even this official history acknowledges that, due to incomplete records, 'it is now impossible to state exactly how many British, Australian or New Zealand troops took part'.
6 McClymont, *To Greece*, 486.
7 Howard Kippenberger, 'Foreword', in W. Wynne Mason, *Official History of New Zealand in the Second World War 1939–45: Prisoners of War* (Wellington: War History Branch, Department of Internal Affairs, 1959), v. Kippenberger's figures for First World War POWs are 25 captured at Gallipoli, 12 in Egypt and 464 in France and Belgium. Recent research indicates the figure could be higher but is likely to remain less than 1000.
8 McGibbon, *New Zealand and the Second World War*, 68; https://nzhistory.govt.nz/war/the-battle-for-crete/overview
9 Harper, *Kippenberger*, 85–86.
10 Antony Beevor, *Crete: The Battle and the Resistance* (London: John Murray, 1991), is probably Freyberg's harshest critic about his use of the Ultra source. Beevor repeats his criticism of Freyberg in *The Second World War* (London: Weidenfeld & Nicolson, 2012), 165. See also McGibbon, *New Zealand and Second World War*, 70.
11 D. M. Davin, *Official History of New Zealand in the Second World War 1939–45: Crete* (Wellington: War History Branch, Department of Internal Affairs, 1953), 480.
12 Beevor, *Crete*, 107.
13 McGibbon, *New Zealand and the Second World War*, 73.
14 Davin, *Crete*, 340.
15 Figures are from Davin, *Crete*, 521, 486. Wounded on Crete, Howard Kippenberger, 'Foreword', in Mason, *Prisoners of War*, v–vi.
16 Beevor, *The Second World War*, 173. See also Davin, *Crete*, 81, 464.
17 Quoted in Harper, *Kippenberger*, 84.

Chapter 4: The Air War over Europe

1 Richard Overy, 'Introduction', in Sebastian Cox and Peter Gray, *Air Power History: Turning Points from Kitty Hawk to Kosovo* (London: Frank Cass, 2002), ix.
2 J. M. S. Ross, *Official History of New Zealand in the Second World War 1939–45: Royal New Zealand Air Force* (Wellington: War History Branch, Department of Internal Affairs, 1955), 39. Ross lists the strength of the RNZAF in September 1939 as: Regular Air Force; 91 officers, 665 airmen; Territorial Air Force: 79 officers, 325 airmen.
3 Paul A. Harrison, 'Royal New Zealand Air Force', in Ian McGibbon (ed.), *The Oxford Companion to New Zealand Military History* (Auckland: Oxford University Press, 2000), 460.
4 Harrison, 'Royal New Zealand Air Force', 461.
5 Christina Goulter, 'Airmen, New Zealand in the RAF in the Second World War', in McGibbon, *Oxford Companion to New Zealand Military History*, 15.

6 Max Lambert, *Night After Night: New Zealanders in Bomber Command* (Auckland: HarperCollins, 2005), 75.
7 Lambert, *Night After Night,* 34–35, 442.
8 Vincent Orange, 'New Zealand Airmen over Europe', in John Crawford (ed.), *Kia Kaha: New Zealand in the Second World War* (Melbourne: Oxford University Press, 2002), 77.
9 Adam Claasen, *Dogfight: The Battle of Britain* (Auckland: Exisle, 2012), 187.
10 Christina Goulter, 'RAF Coastal Command's Anti-shipping Campaign', in Crawford, *Kia Kaha,* 81.
11 Goulter, 'Airmen', 15.

Chapter 5: The War in North Africa

1 John Connell, 'Wavell's 30,000', in Sir Basil Liddell Hart (ed.), *History of the Second World War* (London: Phoebus, 1980), 91.
2 Glyn Harper, *The Battle for North Africa: El Alamein and the Turning Point for World War II* (Bloomington: Indiana University Press, 2017), 8.
3 Harper, *The Battle for North Africa,* 17.
4 Ian McGibbon, *New Zealand and the Second World War: The People, the Battles and the Legacy* (Auckland: Hodder Moa Beckett, 2004), 117.
5 McGibbon, *New Zealand and the Second World War,* 119.
6 Harper, *The Battle for North Africa,* 245.
7 Williamson Murray and Allan R. Millett, *A War to Be Won: Fighting the Second World War* (Cambridge, Mass: Harvard University Press, 2001), 300.
8 W. G. Stevens, *Official History of New Zealand in the Second World War: Bardia to Enfidaville* (Wellington: War History Branch, 1962), 153.
9 Murray and Millett, *A War to Be Won,* 301.
10 Antony Beevor, *The Second World War* (London: Weidenfeld & Nicolson, 2012), 416.
11 Harper, *The Battle for North Africa,* 250–53.
12 Stevens, *Bardia to Enfidaville,* 383.

Chapter 6: The War at Sea

1 Peter Dennerly, 'The Royal New Zealand Navy', in John Crawford (ed.), *Kia Kaha: New Zealand in the Second World War* (Melbourne: Oxford University Press, 2002), 108.
2 Alison Parr (ed.), *The Big Show: New Zealanders, D-Day and the War in Europe* (Auckland: Auckland University Press, 2006), 2.
3 S. D. Waters, *Official History of New Zealand in the Second World War 1939–45: The Royal New Zealand Navy* (Wellington: War History Branch, Department of Internal Affairs, 1956), 424–25.
4 Richard Jackson, 'Fleet Air Arm', in Ian McGibbon (ed.), *The Oxford Companion to New Zealand Military History* (Auckland: Oxford University Press, 2000), 176.
5 Ian McGibbon, *New Zealand and the Second World War: The People, the Battles and the Legacy* (Auckland: Hodder Moa Beckett, 2004), 203.
6 Jackson, 'Fleet Air Arm', 176.
7 Dennerly, 'The Royal New Zealand Navy', 110.
8 'Richardson, Lieutenant-Commander Archibald', in McGibbon, *Oxford Companion to New Zealand Military History,* 449.
9 Denis Fairfax, 'Royal Navy in the Second World War, New Zealanders in', in McGibbon, *Oxford Companion to New Zealand Military History,* 459.
10 Waters, *The Royal New Zealand Navy,* 127.
11 Elizabeth Cox, 'Women in the armed forces', in McGibbon, *Oxford Companion to New Zealand Military History,* 620.
12 Neill Atkinson, *Hell or High Water: New Zealand Merchant Seafarers Remember the War* (Auckland: HarperCollins, 2009), 20.
13 Atkinson, *Hell or High Water,* 38–41.
14 Waters, *The Royal New Zealand Navy,* 168–69.
15 Waters, *The Royal New Zealand Navy,* 266–67.
16 https://nzhistory.govt.nz/war/royal-new-zealand-navy/second-world-war
17 Dennerly, 'The Royal New Zealand Navy', 107.
18 https://www.aucklandmuseum.com/war-memorial/online-cenotaph/record/C26254?n=henry%20Honeyfield&ordinal=0&from=%2Fwar-memorial%2Fonline-cenotaph%2Fsearch

19 Atkinson, *Hell or High Water*, 39, 21.
20 Dennerly, 'The Royal New Zealand Navy', 121.

Chapter 7: The Air War over the Pacific

1 Paul A. Harrison, 'Royal New Zealand Air Force', in Ian McGibbon (ed.), *The Oxford Companion to New Zealand Military History* (Auckland: Oxford University Press, 2000), 461.
2 Ian McGibbon, *New Zealand and the Second World War: The People, the Battles and the Legacy* (Auckland: Hodder Moa Beckett), 85.
3 J. M. S. Ross, *Official History of New Zealand in the Second World War: Royal New Zealand Air Force* (Wellington: War History Branch, 1955), 177.
4 Ross, *Royal New Zealand Air Force*, 312, 314.
5 Harrison, 'Royal New Zealand Air Force', 461.
6 Ross, *Royal New Zealand Air Force*, vii.

Chapter 8: The Ground War in the Pacific

1 W. David McIntyre, *New Zealand Prepares for War* (Christchurch: Canterbury University Press, 1988), 249.
2 See Glyn Harper, 'Threat Perception and Politics: The Deployment of Australian and New Zealand Ground Forces in the Second World War', *Journal of the Australian War Memorial* (April 1992): 36–43.
3 Ian McGibbon, *New Zealand and the Second World War: The People, the Battles and the Legacy* (Auckland: Hodder Moa Beckett), 141.
4 Oliver A. Gillespie, *Official History of New Zealand in the Second World War 1939–45: The Pacific* (Wellington: War History Branch, Department of Internal Affairs, 1952), 144.
5 Gillespie, *The Pacific*, 158.
6 McIntyre, *New Zealand Prepares for War*, 249.
7 McGibbon, *New Zealand and the Second World War*, 142.
8 Gillespie, *The Pacific*, 188.
9 Ian McGibbon, 'Introduction', in Megan Hutching (ed.), *Against the Rising Sun: New Zealanders Remember the Pacific War* (Auckland: HarperCollins, 2006), 24.
10 F. L. W. Wood, *The New Zealand People at War: Political and External Affairs* (Wellington: War History Branch, Department of Internal Affairs, 1959), 277–78.
11 War Cabinet Minute, 11 September 1944, in *Official History of New Zealand in the Second World War 1939–41: Documents Relating to New Zealand's Participation in the Second World War 1939–45*, Vol. III (Wellington: War History Branch, Department of Internal Affairs, 1963), 456.
12 Gillespie, *The Pacific*, 123.
13 McGibbon, *New Zealand and the Second World War*, 142, 143.
14 J. M. S. Ross, *Official History of New Zealand in the Second World War: Royal New Zealand Air Force* (Wellington: War History Branch, 1955), vii.
15 Gillespie, *The Pacific*, 324, 325.

Chapter 9: The Italian Campaign

1 Antony Beevor, *The Second World War* (London: Weidenfeld & Nicolson, 2012), 501.
2 Max Hastings, *All Hell Let Loose: The World at War 1939–45* (London: HarperPress, 2011), 451.
3 Hastings, *All Hell Let Loose*, 453, 454.
4 Major-General Sir Howard Kippenberger, KBE, CB, DSO, *Infantry Brigadier* (London: Oxford University Press, 1949), 344.
5 Ian McGibbon, *New Zealand and the Second World War: The People, the Battles and the Legacy* (Auckland: Hodder Moa Beckett), 169.
6 Glyn Harper and John Tonkin-Covell, *The Battles of Monte Cassino: The Campaign and Its Controversies* (Auckland: Allen & Unwin, 2013), 67–69.
7 Harper and Tonkin-Covell, *Battles of Monte Cassino*, 19.
8 Harper and Tonkin-Covell, *Battles of Monte Cassino*, 20–22; McGibbon, *New Zealand and the Second World War*, 173.
9 Ian McGibbon (ed.), *The Oxford Companion to New Zealand Military History* (Auckland: Oxford University Press, 2000), 251.
10 Hastings, *All Hell Let Loose*, 530.
11 McGibbon, *New Zealand and the Second World War*, 175.

12 John Tonkin-Covell, 'Lieutenant General Bernard Freyberg: A Necessary Commander?', in Glyn Harper and Joel Hayward (eds), *Born to Lead? Portraits of New Zealand Commanders* (Auckland: Exisle, 2003), 118.
13 Glyn Harper, 'Release from Tyranny', https://www.noted.co.nz/archive/archive-listener-nz-2015/release-from-tyranny
14 Hastings, *All Hell Let Loose,* 441.
15 McGibbon, *New Zealand and the Second World War*,. 177.
16 McGibbon, *New Zealand and the Second World War,* 177.

Chapter 10: Prisoners of War

1 W. Wynne Mason, *Official History of New Zealand in the Second World War 1939–45: Prisoners of War* (Wellington: War History Branch, Department of Internal Affairs, 1959), 524–25.
2 Howard Kippenberger, 'Foreword', in Mason, *Prisoners of War*, v–vi.
3 Although airmen fell into enemy hands throughout the war, they made up just 6 per cent of the New Zealanders in captivity. Mason, *Prisoners of War*, 525.
4 Kippenberger, 'Foreword', in Mason, *Prisoners of War*, v. Naval men and merchant seaman made up less than 1 per cent of the New Zealand prisoners of war. Mason, *Prisoners of War*, 525.
5 See Mason, *Prisoners of War*, 160.
6 Ian McGibbon, *New Zealand and the Second World War: The People, the Battles and the Legacy* (Auckland: Hodder Moa Beckett), 209.
7 The best known was Russell Braddon's *The Naked Island*, published in 1950. By 1968 it had been reprinted 11 times and sold a million copies in Britain alone. Nevil Shute's novel *A Town Like Alice*, also published in 1950, became a bestseller and was made into a film and later a television mini-series. It includes an incident where an Australian soldier is crucified for stealing food to give to starving women prisoners. Much more recently, Richard Flanagan's Booker Prize-winning *The Narrow Road to the Deep North* (Melbourne: Vintage, 2013) dealt memorably with the same subject.
8 McGibbon, *New Zealand and the Second World War*, 189.
9 Mason, *Prisoners of War*, ix.
10 Glyn Harper and Colin Richardson, *Acts of Valour: The History of the Victoria Cross and New Zealand* (Auckland: HarperCollins, 2016), 286.
11 https://www.nzherald.co.nz/nz/news/article.cfm?c_id=1&objectid=3051908
12 These are all recorded in Kenneth Sandford, *Mark of the Lion: The Story of Capt. Charles Upham, V.C. and Bar* (London: Hutchinson, 1962).
13 Mason, *Prisoners of War*, ix–x.
14 Mason, *Prisoners of War*, 158.
15 Mason, *Prisoners of War*, 158.

Chapter 11: The Home Front and Jayforce

1 https://www.newworldencyclopedia.org/entry/Total_war
2 Max Hastings, *All Hell Let Loose: The World at War 1939–45* (London: HarperPress, 2011), xx.
3 Ian McGibbon, *New Zealand and the Second World War: The People, the Battles and the Legacy* (Auckland: Hodder Moa Beckett), 149.
4 J. V. T. Baker, *Official History of New Zealand in the Second World War 1939–45: War Economy* (Wellington, Historical Publications Branch, 1965), 92.
5 *Otago Daily Times*, 31 December 1941, 4.
6 McGibbon, *New Zealand and the Second World War*, 102.
7 McGibbon, *New Zealand and the Second World War*, 209.
8 McGibbon, *New Zealand and the Second World War*, 221.

Chapter 12: Coming Home

1 https://www.orwellfoundation.com/the-orwell-foundation/orwell/essays-and-other-works/the-lion-and-the-unicorn-socialism-and-the-english-genius/. The quote is in Part III, II 'The English Revolution'.
2 Nancy M. Taylor, *Official History of New Zealand in the Second World War 1939–45: The Home Front Volume II* (Wellington: Historical Publications Branch, 1986), 1295.

3 Ian McGibbon, *New Zealand and the Second World War: The People, the Battles and the Legacy* (Auckland: Hodder Moa Beckett), 212.
4 This is the theme of Alison Parr's book *Silent Casualties: New Zealand's Unspoken Legacy of the Second World War* (Auckland: Tandem Press, 1995).
5 https://nzhistory.govt.nz/war/second-world-war/counting-the-cost
6 McGibbon, *New Zealand and the Second World War*, 211.
7 John Laffin's controversial book *British Butchers and Bunglers of World War One* was reprinted in 1997 for the fourth time by Sutton Publishing.
8 Max Hastings, *All Hell Let Loose: The World at War 1939–45* (London: HarperPress, 2011), 674–75.

Abbreviations

DAK	Deutsches Afrika Korps
DFC	Distinguished Flying Cross
DFM	Distinguished Flying Medal
DSC	Distinguished Service Cross
DSM	Distinguished Service Medal
DSO	Distinguished Service Order
EATS	Empire Air Training Scheme
FAA	Fleet Air Arm
FAFAI	Forces Available for Anti-Invasion scheme
HMNZS	His Majesty's New Zealand Ship
HMS	His Majesty's Ship
LRDG	Long Range Desert Group
MDS	main dressing station
MID	mentioned in dispatches
NCOs	non-commissioned officers
MV	motor vessel (ship)
NZANS	New Zealand Army Nursing Service
NZEF	New Zealand Expeditionary Force
PG	prigione di guerra (prisoner of war)
RAF	Royal Air Force
RAP	regimental aid post
RNZAF	Royal New Zealand Air Force
RNZN	Royal New Zealand Navy
RNZNVR	Royal New Zealand Navy Volunteer Reserve
SS	steam ship
VC	Victoria Cross
VJ Day	Victory over Japan Day
WAAF	Women's Auxiliary Air Force
WAAC	Women's Auxiliary Army Corps
WRNZNS	Women's Royal New Zealand Naval Service (Wrens)
WWSA	Women's War Service Auxiliary
YMCA	Young Men's Christian Association

Bibliography

Atkinson, Neill (ed.). *Hell or High Water: New Zealand Merchant Seafarers Remember the War*. Auckland: HarperCollins, 2009.

Baker, J. V. T. *Official History of New Zealand in the Second World War 1939–45: War Economy*. Wellington: Historical Publications Branch, 1965.

Barber, Laurie. *War Memorial: A Chronology of New Zealand and World War II*. Auckland: Heinemann Reed, 1989.

——— and John Tonkin-Covell. *Freyberg: Churchill's Salamander*. Auckland: Century Hutchinson, 1989.

Beevor, Antony. *The Second World War*. London: Weidenfeld & Nicolson, 2012.

——— *Crete: The Battle and the Resistance*. London: John Murray, 1991.

Belich, James. *Paradise Reforged: A History of the New Zealanders: From the 1880s to the Year 2000*. Auckland: Allen Lane, 2001.

Claasen, Adam. *Dogfight: The Battle of Britain*. Auckland: Exisle, 2012.

Clifton, George. *The Happy Hunted*. London: Cassell and Co., 1952.

Cox, Sebastian and Peter Gray. *Air Power History: Turning Points from Kitty Hawk to Kosovo*. London: Frank Cass, 2002.

Crawford, John (ed.). *Kia Kaha: New Zealand in the Second World War*. Melbourne: Oxford University Press, 2002.

Davin, D. M. *Official History of New Zealand in the Second World War 1939–45: Crete*. Wellington: War History Branch, Department of Internal Affairs, 1953.

Ell, Gordon. *New Zealand Traditions and Folklore*. Auckland: Bush Press, 1994.

Filer, David. *Crete: Death from the Skies: New Zealand's Role in the Loss of Crete*. Auckland: David Bateman, 2010.

——— *Home and Away: Images of New Zealanders in World War II*. Auckland: David Bateman, 1990.

——— *Kiwis in Khaki: A Pictorial History of the New Zealand Soldier in World War II*. Auckland: David Bateman, 1989.

Gillespie, Oliver A. *Official History of New Zealand in the Second World War 1939–45: The Pacific*. Wellington: War History Branch, Department of Internal Affairs, 1952.

Harper, Glyn. *Kippenberger: An Inspired New Zealand Commander*. Auckland: HarperCollins, 1997.

——— *The Battle for North Africa: El Alamein and the Turning Point for World War II*. Auckland: Massey University Press, 2015.

——— and Colin Richardson. *Acts of Valour: The History of the Victoria Cross and New Zealand*. Auckland: HarperCollins, 2016.

——— and Joel Hayward. *Born to Lead? Portraits of New Zealand Commanders*. Auckland: Exisle, 2003.

——— and John Tonkin-Covell. *The Battles of Monte Cassino: The Campaign and its Controversies*. Auckland: Allen & Unwin, 2013.

Hastings, Max. *All Hell Let Loose: The World at War 1939–45*. London: HarperPress, 2011.

Holmes, Richard. *World War II in Photographs*. London: Carlton Books, 2000.

Hutching, Megan (ed.). *Against the Rising Sun: New Zealanders Remember the Pacific War*. Auckland: HarperCollins, 2006.

Keegan, John. *The Second World War*. London: Arrow Books, 1990.

King, Michael. *The Penguin History of New Zealand*. Auckland: Penguin, 2003.

——— *New Zealanders at War*. Auckland: Heinemann, 1981.

Kippenberger, Major-General Sir Howard, KBE, CB, DSO. *Infantry Brigadier*. London: Oxford University Press, 1949.

Kippenberger, Howard. 'The New Zealand Army'. *Journal of the Royal United Services Institute* 102, no. 605 (1957): 66–74.

Lambert, Max. *Night After Night: New Zealanders in Bomber Command*. Auckland: HarperCollins, 2005.

——— *Day After Day: New Zealanders in Fighter Command*. Auckland: HarperCollins, 2011.

——— *Victory: New Zealand Airmen and the Fall of Germany*. Auckland: HarperCollins, 2014.

Liddell Hart, Basil (ed.). *History of the Second World War.* London: Phoebus, 1980.
Mason, W. Wynne. *Official History of New Zealand in the Second World War 1939–45: Prisoners of War.* Wellington: War History Branch, Department of Internal Affairs, 1959.
McClymont, W. G. *Official History of New Zealand in the Second World War 1939–45: To Greece.* Wellington: War History Branch, Department of Internal Affairs, 1959.
McGibbon, Ian. *New Zealand and the Second World War: The People, the Battles and the Legacy*. Auckland: Hodder Moa Beckett, 2004.
——— (ed.). *The Oxford Companion to New Zealand Military History*. Auckland: Oxford University Press, 2000.
McIntyre, W. David. *New Zealand Prepares for War*. Christchurch: Canterbury University Press, 1988.
McLeod, John. *Myth and Reality: The New Zealand Soldier in World War II.* Auckland: Reed Methuen, 1986.
Murray, Williamson and Allan R. Millett. *A War to Be Won: Fighting the Second World War*. Cambridge, Mass.: Belknap of Harvard University Press, 2000.
Neave, Airey. *They Have Their Exits*. London: Pan Books, 1955.
Official History of New Zealand in the Second World War 1939–41: Documents Relating to New Zealand's Participation in the Second World War 1939–45, 3 vols. Wellington: War History Branch, Department of Internal Affairs, 1949–63.
Orange, Vincent. *Ensor's Endeavour: A Biography of Wing Commander Mick Ensor, DSO & Bar, DFC & Bar, AFRC, RNZAF & RAF.* London: Grub Street, 1994.
Overy, Richard. *Why the Allies Won*. London: Pimlico, 1995.
——— *Bomber Command 1939–45*. London: HarperCollins, 1997.
Parr, Alison. *Silent Casualties: New Zealand's Unspoken Legacy of the Second World War*. Auckland: Tandem Press, 1995.
——— (ed.). *The Big Show: New Zealanders, D-Day and the War in Europe*. Auckland: Auckland University Press, 2006.
Reid, P. R. *Colditz: The Colditz Story*. London: Hodder & Stoughton, 1965.
Rogers, Anna. *While You're Away: New Zealand Nurses at War 1899–1948*. Auckland: Auckland University Press, 2003.
Ross, J. M. S. *Official History of New Zealand in the Second World War1939–45: Royal New Zealand Air Force.* Wellington: War History Branch, Department of Internal Affairs, 1955.
Sandford, Kenneth. *Mark of the Lion: The Story of Capt. Charles Upham, V.C. and Bar*. London: Hutchinson, 1962.
Simpson, Tony. *Operation Mercury: The Battle for Crete, 1941*. Auckland: Hodder & Stoughton, 1981.
Stevens, W. G. *Official History of New Zealand in the Second World War 1939–45: Bardia to Enfidaville.* Wellington: War History Branch, Department of Internal Affairs, 1962.
Taylor, Nancy M. *Official History of New Zealand in the Second World War 1939–45: The Home Front Volume II.* Wellington: Historical Publications Branch, 1986.
Thompson, H. L. *Official History of New Zealand in the Second World War 1939–45: New Zealanders with the Royal Air Force. Volume 1: European Theatre September 1939–December 1942.* Wellington: War History Branch, Department of Internal Affairs, 1953.
Waters, S. D. *Official History of New Zealand in the Second World War 1939–45: The Royal New Zealand Navy.* Wellington: War History Branch, Department of Internal Affairs, 1956.
Wood, F. L. W. *Official History of New Zealand in the Second World War 1939–45: The New Zealand People at War: Political and External Affairs.* Wellington: War History Branch, Department of Internal Affairs, 1958.

Acknowledgements

Writing any book is a team effort, and working on a book such as this based on photographs makes the teamwork even more important. Most of the images came from the archives of the three service museums. We would like to thank Dolores Ho from the National Army Museum at Waiouru, Simon Moody and Matthew O'Sullivan from the Air Force Museum of New Zealand at Wigram, and Rachel Stallard and Michael Wynd from the Torpedo Bay Navy Museum at Devonport, which is the official museum of the New Zealand Navy.

It was surprising to find many photos in the smaller archives around the country. Our thanks go to Lesley Courtney at Manawatū Heritage, Mark Compain at Royal New Zealand Returned and Services Association, Clas Chamberlain and Jo Hollander at the Engineer Corps Memorial Centre at Linton Military Camp, and Neil Frances at Wairarapa Archive. Many of these people volunteer their time in their respective archives.

Our thanks also go to the following individuals who sent in photographs: P. J. Brooke-Taylor, C. R. Lees, Paul Lumsden, Allan Muntz, Tony Goodwin, who is an old friend of O. E. B. Turley, Ian Collins for Ivan Collins and Stephen Parsons. The photos from these donors were from their fathers or friends. We would also like to thank the many others who sent us photos which, we regret, we were unable to use.

We would like to acknowledge and thank Massey University for supporting this project and for awarding it a grant from the Massey University Research Fund.

Fortunately we were able to source all the photos that we needed to before Covid-19 hit and we went into Level 4 lockdown. This enforced period at home gave us the time to finish the book much earlier than planned. We sorted through the photos and made the hard decisions about which photos to use. There were many interesting photos that we couldn't use as there wasn't enough information available to identify where or when they were taken. Glyn also wrote the captions, which Susan, laboriously, typed up. As usual when we're working together there were some 'warm discussions' regarding the content of the book.

This book wouldn't have been possible without the support of Massey University Press; in particular Nicola Legat who commissioned the book and Anna Bowbyes and Tracey Borgfeldt who tied it all together. A special thank you goes to Anna Rogers, a talented author in her own right, who, as usual, did a great job of editing the manuscript. Her suggestions made the book stronger.

We would also like to thank Colonel (Rtd) Colin Richardson, who did a thorough check of the captions in the manuscript.

Although we worked alongside each other for the entirety of this project, Glyn, as the historian, takes responsibility for any errors or omissions.

Glyn Harper and Susan Lemish
Foxton Beach
April 2021

About the Authors

Glyn Harper is Professor of War Studies at Massey University in Palmerston North. Graduating with a master's degree in modern history, and a diploma of teaching from the University of Canterbury, Glyn moved to Australia where he was a high school teacher for seven years. In 1988, Glyn joined the Australian Army and later completed his PhD on the New Zealand commander Sir Howard Kippenberger at the University of New England. In 1996, Glyn transferred to the New Zealand Army where, as commanding officer of the Military Studies Institute, he rose to the rank of Lieutenant Colonel. Glyn was the Army's official historian for the deployment to East Timor and is the best-selling author of more than 20 books for adults. He is also an award-winning author of books for children and young people.

After retiring from the Army in 2001, Glyn joined Massey University. He was Director of the Centre for Defence and Security Studies for six years and was instrumental in establishing the Centenary History Programme. He was Massey's Project Manager for the CHP and wrote two of its 14 volumes. In 2010, Glyn was a Fulbright Senior Scholar in the United States and in 2012 he was awarded a Queen's Service Medal for services to historical research.

Susan Lemish, originally from Melbourne, moved to New Zealand in 2001. For the past 20 years she has assisted Glyn Harper with the research and writing of many books.

Index

Index of military units and formations, including ships

The publisher wishes to thank the Air Force Museum of New Zealand, the National Army Museum Te Mata Toa, and the National Museum of the Royal New Zealand Navy Te Waka Huia o te Taua Moana o Aotearoa for their help in making this book possible.

MASSEY
UNIVERSITY
PRESS

First published in 2021 by Massey University Press
Private Bag 102904, North Shore Mail Centre
Auckland 0745, New Zealand
www.masseypress.ac.nz

Text copyright © Glyn Harper and Susan Lemish, 2021
Images copyright © as credited, 2021

Design by Alice Bell
Cover photographs: (front, top) New Zealand machine gunners carry their equipment during a field exercise. National Army Museum, 2016-339-1; (back, top) An unknown pilot of No. 260 Squadron gives the V for victory sign from the cockpit of his Hurricane. Air Force Museum of New Zealand, ALB863701094

The moral right of the author has been asserted

All rights reserved. Except as provided by the Copyright Act 1994, no part of this book may be reproduced, stored in or introduced into a retrieval system or transmitted in any form or by any means (electronic, mechanical, photocopying, recording or otherwise) without the prior written permission of both the copyright owner(s) and the publisher.

A catalogue record for this book is available from the National Library of New Zealand

Printed and bound in China by Everbest Investment Ltd

ISBN: 978-0-9951407-3-8